chapter 3

0 6 - 1g.

Experiment #2

wt. of test tube = 20.89

f v . .7679

T.L.C plate (SO$_2$)

PREP(.5mm)

17g/ 34ml H$_2$O)   ANALYTICAL (.25mm)

33g /58 ml H$_2$O   PREP (.5)

Chromic Acid

5g /100ml   (100g NaCr$_2$O$_7$ / 1 jug of tech. conc H$_2$SO$_4$

W. B. SAUNDERS COMPANY
PHILADELPHIA • LONDON • TORONTO

1971

SAUNDERS GOLDEN SERIES

EXPERIMENTAL METHODS IN ORGANIC CHEMISTRY

JAMES A. MOORE

DAVID L. DALRYMPLE

Department of Chemistry, University of Delaware

W. B. Saunders Company: West Washington Square
Philadelphia, Pa. 19105

12 Dyott Street
London WC1A   1DB

1835 Yonge Street
Toronto 7, Ontario

Experimental Methods in Organic Chemistry

Print Number: 9   8   7   6   5   4   3   2   1

PREFACE

In developing material for a modern introductory organic laboratory, a number of choices must be made, and a course must be steered between several conflicting factors. Schedules are tight for both students' time and laboratory space and usually set a fairly rigid three-hour limit on the laboratory period. With large groups of students, highly individualized experiments become unmanageable, but the course must not be restricted to a collection of simple cut-and-dried procedures. Rather, the increasing sophistication of experimental organic chemistry requires more elaborate experiments to provide meaningful and useful experience. Students vary greatly in ability, and this must be taken into account in designing experiments from which individuals at both ends of the spectrum can profit. In preparing this text, we have kept these points before us and have sought a balance in which each is recognized and hopefully satisfied as fully as possible.

A major decision concerns the use of instrumental methods. Contemporary laboratory work in organic chemistry depends heavily on chromatographic analysis and spectroscopy. These methods should be introduced in a first course and used in more than a perfunctory way. For most nonmajors, organic lab is the only opportunity to learn something of these methods, whereas they should be among the primary tools of a chemistry major. We have therefore not hesitated to make VPC or TLC analysis of reaction mixtures the main focus of several chapters; these have proven to be among the most successful and popular experiments in the course at the University of Delaware. The use of spectroscopic equipment by the student is called for only in connection with the chapter on unknowns, but spectral data are presented in a number of experiments and are incorporated in the questions.

The experiments are organized in three main groupings. The first 11 chapters cover the techniques and methods that are used throughout the course. These allow time for classroom work to progress to a point where reactions and compounds can be meaningfully studied in the lab. When practical, however, the experiments on separation and purification involve work on simple reaction mixtures obtained by the student. It is much more profitable and interesting to the student to work up a reaction that he has run, even though the scope and mechanism are not completely under-stood, than to work with an anonymous mixture from the side shelf. Several compounds are encountered more than once in these early chapters; this serves to integrate the material and also reduces the number of different chemicals that must be procured and dispensed. Chapters 1 to 7 are organized to be covered in that order.

Chapters 9 and 10 are intended to provide a minimum background in spectral interpretation for work on unknowns; reference to this material can be made at any point that ties in with the coverage of spectra in lecture or classroom schedules. Chapter 11 on the literature is somewhat more comprehensive, since this material is not as readily accessible elsewhere. This chapter includes an exercise in locating compounds and preparative methods in the literature and provides optional experiments if in-dividualized laboratory work is possible.

The sequence of experiments in Chapters 12 to 34 follows that of most contemporary texts, such as Morrison and Boyd's *Organic Chemistry;* aliphatic and aromatic chemistry are integrated and amines and more specialized topics such as heterocyclic compounds are covered relatively late. These chapters need not, however, be taken up in any particular order. The experiments have been designed to provide a variety of perspectives. In some cases, the thrust is primarily synthetic; in others it is the analysis of a complex product mixture or a comparison of reactivity. A number of these chapters contain some element of a problem that must be answered by experimental data, thereby emphasizing the importance and sig-nificance of the student's observations.

The final chapter on unknowns is a rather ambitious one, but we have had good results and enthusiastic reception in both major and nonmajor courses. In this approach to identification, primary emphasis is placed on infrared and nmr spectra, as it must be in any modern curriculum. Some of the time-honored classification tests are indisputably valuable in learn-ing the reactivity of functional groups. These have been retained, in this context, in earlier chapters dealing with functional groups rather than using them as identification tools of uncertain value.

This experiment is intended for a minimum of 6 to 8 laboratory periods and can be extended to a longer sequence, depending on the number and difficulty of the unknowns. We have had students obtain their own infrared spectra; this has presented no difficulties, although with large sections, spectra may have to be run in advance of the experiment or in extra periods. An experiment for large sophomore courses involving nmr spectra of hundreds of unknowns poses a severe logistical problem, however. Our approach to this problem was to develop a list of some 200 compounds suitable for unknowns, record the spectra, and provide the student with a copy of the spectrum with his sample of the unknown. Through the cooperation of the publishers, a set of these spectra is available, on request, to teachers who adopt this text. Additional sets may be purchased from the publisher. Authorization to reproduce additional copies of individual spectra for student use is also granted.

Finally, a word about the scope and length of this text. One approach to a book of this type is to attempt to include as much material as possible that is relevant to the experiments and make the book as fully self-contained as possible. We have held the view that the typical student will read only as much of the material in a laboratory text as is needed for the experiment at hand, and we have attempted to restrict the text to this extent. For several topics, particularly in the work on unknowns, other sources must be consulted.

We wish to thank the groups of students who have refined the experiments and provided the original incentive to prepare this book, and the graduate instructors for their assistance and helpful comments. We express our appreciation also to Professor Arnold Krubsack of Ohio State University, Professor Richard Taber of Colorado College, Professor Grant Taylor of the University of Louisville, and particularly Professor James A. Garrison of the State University of New York at Buffalo and Professor Charles Scanio of Iowa State University for their careful readings and constructive criticism of the manuscript. The expert and patient preparation of the manuscript in several versions by Mrs. Mary Hinton is most gratefully acknowledged.

JAMES A. MOORE
DAVID L. DALRYMPLE

Newark, Delaware

CONTENTS

INTRODUCTION

LABORATORY SAFETY

Most organic compounds are flammable and some have irritating or toxic vapors; many organic reactions are potentially violent. Some specific hazards are mentioned in the next few pages and throughout this book when they arise. Accidents are minimized by good sense and observance of some elementary precautions, but they do occur, and you should familiarize yourself with the location of fire extinguishers, fire blankets, safety showers, and eyewash fountain and know how and when to use these devices.

Personal Protection. Two important rules should be observed without exception:

EYE PROTECTION MUST BE WORN AT ALL TIMES IN THE LABORATORY REGARDLESS OF WHAT IS BEING DONE. Lightweight safety glasses with side shields are available at low cost, and are most satisfactory. Prescription glasses (*not* sunglasses) are acceptable. Contact lenses provide no protection—safety glasses are required.

NEVER WORK IN A LABORATORY WITHOUT ANOTHER PERSON BEING PRESENT OR WITHIN CALLING DISTANCE. Minor accidents can become disasters if help is not available.

In addition to these general rules, a number of specific points can also be mentioned.

1. Never taste any compound in the laboratory.

2. In determining the odor of a compound, bring the stopper of the bottle cautiously toward the nose; do not inhale.

3. Avoid contact of chemicals with the skin, especially the face; wash hands as soon as possible after making transfers or other manipulations.

4. When inserting glass tubing into a rubber stopper or rubber tubing, lubricate with a drop of glycerine and protect hands with a towel. Thin-wall disposable pipets should not be used as connectors for rubber tubing or stoppers. They are fragile and very easily crushed.

5. Be constantly mindful never to heat a flask or any apparatus that is not opened to the atmosphere.

Fire Hazards. Although electric heating elements have displaced gas flames for many uses, burners are necessary in most undergraduate laboratory work. There is a definite hazard when burners are used, and the following precautions must be observed:

1. Never heat an organic liquid over a flame except under a condenser. When refluxing a liquid, be certain that the condenser is tightly fitted. If a temperature below 95° is sufficient, use a steam bath rather than a burner.

2. Before lighting a flame, check to see that volatile liquids are not being poured or evaporated in your vicinity. Organic vapors are heavier than air and flow downward; they diffuse rapidly and can be ignited by a flame several feet away from the source.

3. Conversely, before pouring or evaporating a volatile liquid, be certain that none of your neighbors is using a flame.

4. Always turn a burner off as soon as you are finished using it—never leave it on unnecessarily.

5. As a general practice, and particularly if a burner is in use, avoid loose-fitting long sleeves and cuffs; long hair should be tied back during laboratory work.

6. Smoking creates an unnecessary fire hazard and is not permitted in the laboratory.

Disposal of Chemicals. Water-immiscible organic solvents and other liquids should be discarded in a designated waste-solvent can. They should never be poured into a sink if at all possible. If only a sink is available, flush thoroughly with water; even then, residues will remain in a trap, and vapors can persist in the sink for some time. Moreover, any chemicals discharged in a drain eventually add to pollution of the environment.

Water-insoluble solids and glass should be disposed of in a non-metallic chemical waste jar; do not throw them into a sink or waste paper basket.

Chemicals that react vigorously with water, such as acid chlorides or alkali metals, should be decomposed in a hood in a suitable way, e.g., reaction with alcohol.

First Aid. For any chemical spilled or splashed on the skin, the first step is to flush liberally with water for several minutes. Use an eyewash fountain if face or eyes are affected, or a sink faucet if an eye wash is not close by. For acid burns, sodium bicarbonate can be applied for first aid. For minor heat burns, cold water will lessen the pain; salves are not recommended. Any serious burn from fire or chemicals should be referred to a physician immediately.

PREPARATION FOR THE LABORATORY

For successful laboratory work it is essential that, before beginning an experiment, you understand what you are going to do, and why and how you are going to do it. Study the assigned chapter in advance and plan your operations. Answer, in writing if required by your instructor, any questions marked with a star (⋆) at the end of the experiment. These questions provide background for the work and will acquaint you with points that can and should be understood before you actually do the experiment.

EQUIPMENT AND OPERATIONS

Experimentation in organic chemistry calls into play a number of operations and techniques and a rather large assortment of apparatus. Detailed instructions in the techniques and equipment used for various separation and purification methods are given in Chapters 2 to 4 in conjunction with actual experimental procedures. This section covers a few points of general practice that apply to nearly all experiments.

Glassware.   A typical set of glassware with standard taper ground joints for use in an undergraduate course is shown in Figure 1.1. This glassware permits you to assemble apparatus quickly and securely; it is expensive and must be handled with care. Other equipment which should be available for the experiments in this book include Erlenmeyer flasks, beakers and test tubes in several sizes, funnels, graduated cylinders, pipets, thermometers, spatulas, and assorted clamps and other hardware.

For effective laboratory work it is most important that you develop good working habits and learn the proper equipment for a given purpose and how to use it. Maintain a well organized locker or equipment drawer;

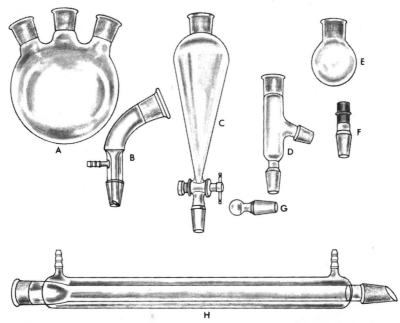

FIGURE 1.1   Standard taper glassware. (*a*) 3-necked round-bottom flask; (*b*) vacuum take-off adapter; (*c*) dropping funnel; (*d*) distillation head; (*e*) round-bottom flask; (*f*) tubing adapter; (*g*) standard taper stopper; (*h*) condenser.

keep your equipment clean and as conveniently located as possible. Make a practice of washing or rinsing glassware as soon as it has been emptied. Take enough time to clean up and store equipment properly at the end of the day.

Reaction Set-ups. Most reactions involve the combination of two reactants at a controlled rate. Many simple reactions on a small scale require only a pipet and test tube, which can be swirled by hand. In other cases, the reactants are combined and heated in a simple set-up, such as that in Figure 1.2. The temperature is controlled by a heating bath or by the boiling point of a reactant or solvent. Mixing, if needed, is provided by the turbulence of boiling. If the reaction is exothermic, cooling occurs by the heat being transferred to the condenser by the refluxing solvent; if necessary, the flask can be immersed in an ice bath.

For reactions on a larger scale or those requiring special conditions, a more elaborate set-up such as that shown in Figure 1.3 may be needed.

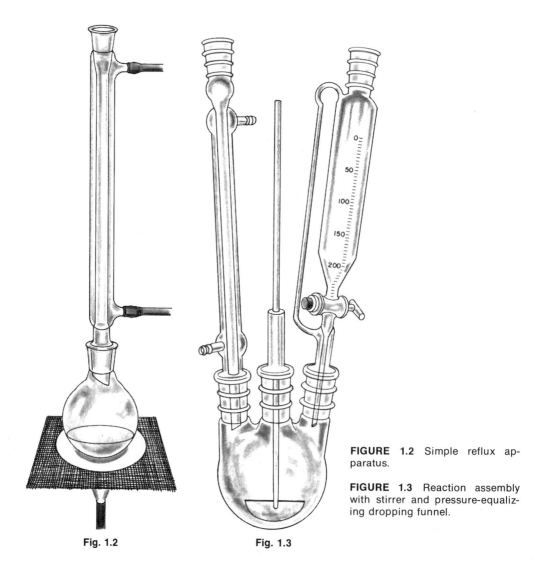

FIGURE 1.2 Simple reflux apparatus.

FIGURE 1.3 Reaction assembly with stirrer and pressure-equalizing dropping funnel.

Fig. 1.2   Fig. 1.3

The separate necks for the dropping funnel, stirrer, and condenser permit flexible assembly and easy control of addition and mixing.

Stirring is important if a reactant must be added and dispersed at a steady rate, or if reaction must occur between separate phases. For relatively small flasks and for reaction mixtures that are not viscous, a bar magnet with an inert coating is placed in the flask and is spun by a motor-driven magnet below the flask. Alternatively, as in Figure 1.3, a rod with a small propeller or paddle, turning in a closely fitting sleeve, is driven by an electric or air-powered motor.

The atmosphere in a reaction may be critical. If the reactants or products are sensitive to water, a drying tube filled with desiccant attached to the condenser may be sufficient. For scrupulous removal of moisture, or when oxygen must be excluded, a dry inert gas, such as nitrogen, is passed into and over the reaction mixture and out through a trap. With this set-up, the dropping funnel must be equipped with a pressure-equalizing arm, as illustrated in Figure 1.3.

Heating. The two sources of heat in most undergraduate labs are the Bunsen burner and steam bath. A burner is a convenient and rapid source of heat; it is also the cause of most laboratory fires, as noted previously. When heating a flask over a burner, place a wire gauze under the flask to distribute the heat, and be ever-vigilant about organic vapors.

When possible, the steam bath (Fig. 3.5) should be used for heating or evaporating any organic liquid. Connect the bath with the steam in the *upper* side arm and condensate in the lower. Place the flask to be heated on the largest ring that will support it. Do not remove the rings and set the flask on the bottom. Turn on steam until it just begins to escape; no benefit is gained from billowing clouds.

Electric heating elements, controlled by a variable transformer, are widely used in a number of forms. The element can be used as a winding or blade to heat a liquid bath, or hot plate, or it is imbedded in a glass-fiber mantle which fits snugly around a flask. Electrically heated devices have replaced burners in many laboratories for all purposes except glass working.

Use of the Aspirator. An aspirator provides a convenient source of reduced pressure for several operations. Water should be turned on to full capacity when an aspirator is used. Splashing may be a problem; this can be met by tying a piece of rag around the outlet or wedging a piece of wire gauze in the trough below the stream. Heavy wall rubber tubing should be used when connecting apparatus to the aspirator since regular tubing will collapse and pinch off the system. If the water pressure drops, or the water is turned off while the aspirator is in use, water tends to be drawn back through the arm into the evacuated system. Most aspirators contain a check valve to prevent this, but a trap should always be connected between the aspirator and the evacuated vessel (Fig. 1.4). The vacuum should always be released *before* turning off the water.

Handling and Measuring Chemicals. In most preparative experiments, solid starting materials and reagents can be weighed on a beam balance

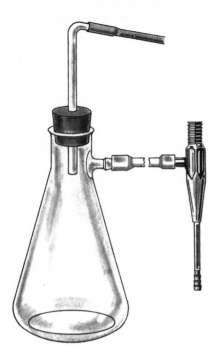

FIGURE 1.4 Aspirator and safety trap.

to ±0.1 g. A top loading automatic balance is a great convenience. A beaker is the most convenient vessel for handling more than a few grams. For weighing small quantities of reagents or products, glassine paper (not filter paper) should be used. The sample can then be placed in a vial or added to a solution simply by picking up the paper by opposite edges or corners and using it as an open funnel. Finely divided solids can be transferred quite completely by gently scraping with a spatula. Metal spatulas should be wiped clean after any use and polished occasionally.

It is unnecessary to weigh certain reagents such as activated carbon, salt, or drying agents; the appropriate amount should be estimated by bulk. Small amounts of sodium or potassium hydroxide need not be weighed. The pellets are fairly uniform and weigh roughly 0.1 g each. If more accurate measurement of a small quantity is required, a solution of known concentration should be used. Never attempt to weigh hydroxide pellets on paper or a watch glass; a few inevitably roll away and liquefy into a highly corrosive puddle.

Liquid reagents and starting materials are conveniently measured by volume in a graduated cylinder or sometimes with a graduated pipet, but these are accurate to only a few per cent. For more accurate measurement liquids should be weighed, particularly if they are rather viscous. For transfer of small volumes of liquids and very approximate measurements, thin-wall, soft-glass transfer pipets are particularly useful. A supply of these should be maintained; they can be rinsed and reused repeatedly. Rubber bulbs of 1 and 2 ml capacity will fill these pipets to about half and full capacity, and there will be many occasions to obtain approximate volumes of liquid in this way. A detailed procedure for transferring small volumes of organic solutions is described in Chapter 3.

Thermometers and Melting Points.   The melting point of a substance is the characteristic temperature at which crystalline solid and liquid are in equilibrium. Since the melting point is easily determined on a very small sample, it is an important and useful physical property.

Determination of the melting point requires a thermometer and a means of heating the sample, in close contact with the thermometer bulb, at a steady controlled rate. This can be carried out either in a liquid bath (Fig. 1.5) or by placing the sample on a metal block in which the thermometer is imbedded (Fig. 1.6). The former has the advantage that each student can calibrate and use his own thermometer and carry out the reading at his own desk. The melting point bath is designed to provide convection and even heating without the need for stirring. The bath contains a high boiling oil and can safely be heated to 250° or higher for brief periods. Heat should be applied with a micro-burner on the side loop; the thermometer bulb and sample are placed just below the upper arm of the loop.

The sample is enclosed in a glass capillary tube which is sealed at one end. Commercially available 2 mm capillaries are somewhat thicker than necessary, but are satisfactory for most purposes. Thinner-walled tubes of smaller diameter can be pulled from a disposable pipet if desired. With either tube, be sure that the end is completely sealed. Introduce the sample by tamping the open end of the capillary in a little pile of crystals, inverting, and shaking down by tapping the bottom on the table top or gently stroking with a file. A 1/8-inch column of solid in the bottom of the tube is plenty. The tube is held onto the thermometer by a thin rubber ring sliced from a piece of rubber tubing. The end of the tube with the sample should be as

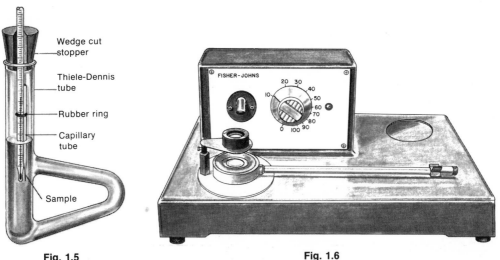

Fig. 1.5

Wedge cut stopper

Thiele-Dennis tube

Rubber ring

Capillary tube

Sample

Fig. 1.6

FISHER - JOHNS

FIGURE 1.5   Thiele tube melting point bath.

FIGURE 1.6   Fisher-Johns melting point block.

close as possible to the thermometer bulb; the ring should be *above* the surface of the oil.

With the melting point block, the sample (less than the amount needed for a capillary melting point) is simply sandwiched between cover glasses which are placed on the block. The rate of heating is controlled by the dial; a setting of 30 will raise the block temperature to about 200°.

With either bath or block, the rate of heating in the vicinity of the melting point should be no more than 3°/min, since heat transfer through glass is slow, and an equilibrium temperature is to be observed. The temperature should be raised more rapidly until it reaches about 15° below the expected melting point. If the range of the melting point is not known, it pays to heat more rapidly and get an approximate value and then repeat at the proper rate.

The observed melting point, as it is usually determined, is actually the range of temperature, normally at least 0.5°, during which the crystals melt entirely. If too large a sample is used, the time required for complete melting is longer, and the temperature range is larger.

For most reasonably pure organic solids, the observed melting point is fairly sharp. A broad range of several degrees is usually an indication of impurity. Very generally, impurities, including solvent, will cause broadening *and lowering* of the melting point, because of the shape of the usual phase diagram for two solids and a solution. A sharp melting point, unchanged by recrystallization, is therefore one criterion of purity. Conversely, this effect can be used as an indication of the identity of two crystalline samples having the same melting point; if they are the same compound, the melting point of an intimate mixture of the two will be undepressed.

A melting point determination is only as accurate as the thermometer used, and before taking an unknown melting point, you should first calibrate the thermometer by checking a few compounds of a known, sharp melting point. For most work in this course, a 250° thermometer is preferable because it has larger graduations. For melting points from 250 to 300°, the Fisher-Johns block should generally be used to avoid darkening of the oil bath. A known melting point in this range should be checked also; no assurance can be offered for the accuracy of the thermometer in the block.

The following compounds are convenient for checking the thermometer:

COMPOUND	MELTING POINT	COMPOUND	MELTING POINT
p-dichlorobenzene	53°	salicylic acid	159°
acetanilide	114°	p-nitrobenzoic acid	239°

SUBMISSION OF SAMPLES AND REPORTS: PERCENTAGE YIELD

In most of the experiments to be carried out, a sample of compound is to be turned in or a specific result may be called for in written form.

Samples should be submitted in an appropriate size plastic cap vial and neatly labeled with the student's name, the name of the product, its weight, percentage yield, melting point (mp) or boiling point (bp), and any other data specified by the instructor.

The *yield* in a reaction is the amount of product of acceptable purity actually obtained (the term should not be used for the product itself). The percentage yield is the ratio of this amount to that theoretically obtainable $\times 100$. In a reaction involving more than one starting material, the theoretical or 100 per cent yield is determined by the reactant which is used in the smallest stoichiometric quantity. In calculating a percentage yield, therefore, one must first determine the molar amounts of each starting compound and then the theoretical amount of product that could be obtained from the limiting reactant.

As an example consider the esterification of succinic acid; 5.0 g of the acid is heated with 100 ml of ethanol and 6.2 g of ester is isolated.

$$HOCOCH_2CH_2CO_2H + 2\ C_2H_5OH \xrightarrow{H^+} C_2H_5OCOCH_2CH_2CO_2C_2H_5 + 2\ H_2O$$

succinic acid ⸱⸱⸱⸱⸱⸱⸱ ethanol ⸱⸱⸱⸱⸱⸱⸱ diethyl succinate

Calculating molar amounts:

succinic acid	ethanol	diethyl succinate
5.0 g; mol wt 118	100 ml; d, 0.79 g/ml	6.2 g; mol wt 174
= 0.042 mole	= 79 g; mol wt 46	= 0.036 mole
	= 1.72 moles	

The ethanol is present in large excess (0.084 mole required) and the yield of ester product is thus based on the acid. The theoretical amount of diethyl succinate is the same as that of acid, 0.042 mole, and the percentage yield is $\frac{.036}{.042} \times 100 = 86\%$.

In the foregoing example, if 0.40 g of unreacted succinic acid were recovered from the reaction in usable form, the yield might alternatively be based on "unrecovered starting material," i.e., $5.0 - 0.4 = 4.6$ g; the percentage yield on this basis would be 92 per cent.

RECORDS

Equally as important as the actual product or the final report, is the LABORATORY NOTEBOOK. An important objective of this course is to develop good practices and habits in keeping permanent notes of experimental work and in working in an orderly systematic way.

The laboratory notebook must be bound with a hard cover and used only for experimental data in this course. It must, of course, be at hand at all times while you are in the laboratory. The notebook is the fundamental

Diels Alder Reaction of Cyclopentadiene and Maleic Anhydride   Sept. 6, 1970

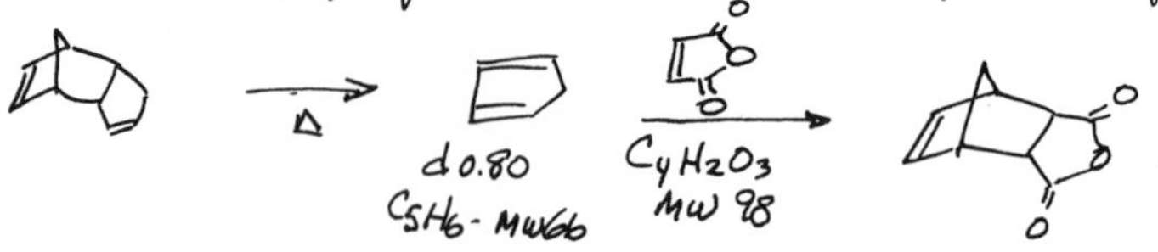

d 0.80
C_5H_6 - MW 66

$C_4H_2O_3$
MW 98

98+66 = 164
MW

Prep calls for:
  20 ml "dicyclopentadiene"
  6 ml cyclopentadiene = 6 × 0.80 = 4.8g → $\frac{4.8}{66}$ = 0.73 moles
  6 g maleic anhydride = 6/98 = 0.061 moles

   20 ml of dicyclopentadiene was placed in 100 ml r.b flask. Fractionating column and condenser attached -- ice-cooled 50 ml r.b used as receiver, with vacuum take-off adapter & $CaCl_2$ tube: Had to distill very slowly to avoid liquid fo thingover. Thermometer fluctuated around 40-42°. Distilled about ⅓ - ½ of total -- stopped 2:15. Distillate pretty clear, but added about 8-10 lumps of $CaCl_2$.

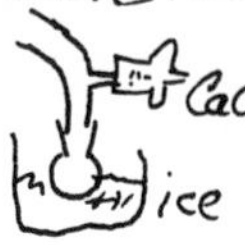

43.6
37.6 tare
―――――
6.0 g
Mal. anhy.

   6.0 g maleic anhydride in 50 ml Erlenmyer. Dissolved in 20 22 ml ethyl acetate (warmed to dissolve, 10ml hexane added; started to xslize again at room temp.

   Added 6.0 ml of the distilled cyclopentadiene (still cold).

4.18
.54
―――――
3.64 g

Soln became perceptibly warm and then tremendous crystallization! Warmed on steam bath to dissolve and cooled slowly -- beautiful long xstls. Collected on Büchner -- washed with a little cold EtOAc-hexane (1:1). Dried in air: 3.64 g of endo-norbornene-5,6-dicarboxylic anhydride, mp 162-164° (lit 164-165°). TLC of ML ($CHCl_3$):   Still more product in ML but not time to isolate.

1	: ○	○	1 xstls mp 162
2	: ∞		2 ML
3	: ○		3 Mal. anhydride

$\frac{3.64}{164}$ = 0.022 mole

$\frac{.022}{.061}$ × 100 =

36 % yield

record of actual laboratory operations and observations. It should provide an account of *what* was done, *how* it was done, and *what happened*. The apparatus used, the sequence of steps, all measurements, significant time intervals, changes in appearance, and other relevant data should be recorded. It should be possible for someone else to repeat the experiment, as you did it, and obtain the same results. The notebook should *not* be:

1. A verbatim transcription of a procedure from the laboratory manual or any other source which you then purport to follow.

2. An *ex post facto* scrapbook of recollections and miscellaneous jottings.

It is impossible to reconstruct a record from isolated numbers and memory; on the other hand, your experimental data cannot be recorded before they exist. It is necessary, therefore, to record the salient operations, measurements, and observations insofar as possible *as they are done or made.* This procedure usually will not result in a flawless copybook record, but it should be coherent and legible.

Each experience should be dated (each day if protracted) and placed on a separate page or pages. It is a good idea to use only right hand pages for the actual write-up; calculations and the like can be recorded on the facing page.

A common objection raised by students is, "Why must I write down in a notebook all of the steps in a procedure which I am following in a lab manual?" The reason for doing this is to provide an orderly account into which your own data and observations can be incorporated. Moreover, it is a habit that must be acquired. In later experiments, you will be adapting a general procedure to your own situation; your own specific case is unique, and the record of what you do is vital.

On page 10 is an illustrative example of an actual notebook page that may serve as a guide to the type of record that should be kept in a typical experiment.

References

Laboratory Safety

N. V. Steere (ed.), *Handbook of Laboratory Safety*. Chemical Rubber Co., Cleveland, 1967.

N. I. Sax, *Dangerous Properties of Industrial Materials*, 3rd Ed., Sec. 12. Van Nostrand Reinhold, New York, 1968.

Apparatus and Techniques—General

A. I. Vogel, *Practical Organic Chemistry*, 3rd Ed., Chapt. II. Longmans, Green and Co., New York, 1956.

K. Wiberg, *Laboratory Technique in Organic Chemistry*. McGraw-Hill, New York, 1960.

2

CRYSTALLIZATION

A compound that is a solid at room temperature is usually isolated and purified by crystallization. A solid may be present in solution as a crude reaction product, or in a fraction from an extraction (Chapt. 3) or chromatogram (Chapt. 7). The substance can be recovered by evaporation of the solution to a dry residue, or by simple precipitation (as in the addition of water to an alcoholic plant extract), but there will be no fractionation or purification. However, in crystallization, or recrystallization, the compound is allowed to separate by a selective process of crystal growth, in which impurities are retained in the mother liquor.

Crystallization depends primarily on solubility relationships. With a few exceptions, the solubility of a compound in any solvent increases markedly with temperature, often manyfold over a temperature range of 40 to 50°. The solubility of an organic solid in an organic solvent usually increases very rapidly as the temperature approaches the melting point of the solid, since two organic liquids of similar type are generally miscible in all proportions.

When a compound has an adequate solubility (a few per cent or more) in a hot solvent, it can often be crystallized quite completely just by cooling. If the solubility is too high at room or ice bath temperature, some of the solvent is removed by evaporation, and a second solvent, of lower solvent power for the desired compound, is added. Ethanol plus water or ether plus hexane are solvent mixtures commonly used for this purpose. In using this approach, care must be taken not to lower the solubility to the point that the compound separates initially as an oil. Even if this subsequently solidifies, purification does not result since impurities will usually have been concentrated in the oil. This difficulty is frequently encountered in using alcohol-water mixtures for recrystallization of low melting compounds.

The solubility of organic solids is a function of the relative polarity of the solvent and solute, and also of the energy of the crystal lattice which must be broken down. "Like dissolves like," and compounds will therefore be most soluble in solvents which are of a similar nature. Compounds in which polar groups (particularly —OH, —NH, —CO_2H, or —CONH—

which can form hydrogen bonds) comprise a major part of the molecule are usually more soluble in hydroxylic solvents such as water or alcohols than in hydrocarbons such as benzene or hexane. Conversely, the latter are suitable solvents for a wide range of compounds of intermediate or low polarity. Regardless of the type of compound, however, the more stable the crystal, as reflected by the melting point, the less soluble it is. This effect may be seen in the melting points and solubilities of the isomeric nitrobenzoic acids. The *para* isomer, having a highly symmetrical structure which can pack into a more stable lattice than the others, has a much higher melting point and correspondingly lower solubility in any given solvent.

Isomer	ortho	meta	para
Mp	147°	141°	242°
Solubility, %			
Ethanol	28	33	2.2
Ether	21	25	0.9

Since crystallization involves the growth of a lattice by addition of like molecules in an ordered way, it can be a highly selective process. In this sense, it is unlike distillation, extraction, or chromatography. These three separation methods depend on the distribution of different molecular species in the two phases strictly according to an equilibrium process. Crystallization may take this form, with a mixture of two solutes crystallizing together in proportion to the amounts present and the solubilities, but this is not always the case. Once the growth of a crystal of one species has begun, like molecules can be added to the lattice until the solubility of that compound has been reached even though the solution may be supersaturated with respect to the second component. If two solutes do crystallize simultaneously as two distinct crystalline phases, fractional crystallization may effect separation. This process involves removing successive crops of differing composition and recrystallizing them, but it is usually far less efficient than chromatography.

STEPS AND TECHNIQUES IN CRYSTALLIZATION

An impure solid to be recrystallized or a gum to be crystallized must first be freed of any impurities that are insoluble in the crystallization solvent. This is done by filtration of a hot subsaturated solution prior to crystallization by cooling; this operation must be carried out rapidly to avoid premature crystallization. Filtration through paper in a funnel with a wide stem or with no stem is the simplest procedure. The solution is collected in an Erlenmeyer flask or test tube; these are the proper vessels for the crystallization process. It is preferable to fold the paper in a fluted form to provide maximum surface area.

To prepare a fluted paper, lay a 4-inch disc of filter paper on the bench-top, and fold it in half and then into quarters. Reopen the latter fold like a book (Fig. 2.1a) and fold corners A and C up to meet point B. Reopen to a semicircle (Fig. 2.1b) and fold corner A first to point D and then to point E, and corner C to D and E, reopening to a semicircle after each fold. The paper should now look like Figure 2.1c, with the folds bending the semicircle into a partially open cone. In each of the eight segments make a fold in the center, alternating with and in the opposite direction to the previous folds, in accordion fashion, to give a fan-like arrangement. When opened at the first fold (AC), a fluted paper (Fig. 2.1d) results.

A more convenient way to remove a granular solid (e.g., drying agent) or small amounts of solid debris is by filtration through a small plug of cotton. Stuff a pea-sized wad of cotton loosely in the bottom of a glass funnel and wedge it slightly into the top of the stem with the tip of a spatula. This method is fast and eliminates the need to rinse a large piece of paper. For filtering a few ml of solution, a soft glass transfer pipet can function as a very convenient microfunnel. With the tip of a second pipet, a small pellet of cotton is simply pushed through from the top and wedged gently against the constriction above the tip. Snip off the tip to a length of about 2 cm and transfer the solution to be filtered with a pipet.

If finely divided solids are present, it may be necessary to clarify the solution by filtration with reduced pressure, using a Büchner or Hirsch funnel (see following discussion). In this case a layer of Celite, which is a finely divided mineral filter aid, is placed on the paper and moistened with solvent. This retains fine particles and permits a fresh surface to be exposed by scraping if filtration becomes too slow. If charcoal is removed by suction filtration, Celite should always be used. When a solution that is near its boiling point is filtered under reduced pressure, clogging of the funnel and bumping or frothing of the filtrate may occur due to rapid evaporation of the solvent.

Besides insoluble impurities, other contaminants may sometimes be removed from the solution in this clarification step by adding activated charcoal before filtration. The higher-molecular-weight compounds, which are usually responsible for dark color in solutions of organic compounds, are selectively adsorbed on the activated carbon. The use of carbon is by no means a panacea for removing impurities; it may be ineffective in removing color, and some loss of the desired solute can occur. If the compound to be crystallized is expected to be colorless and the solution is quite dark, it pays to try the effect of carbon on a small test portion. The decolorizing action of carbon is generally much more efficient in aqueous solutions than in nonpolar organic solvents.

The clarified solution is either cooled, or if necessary concentrated or

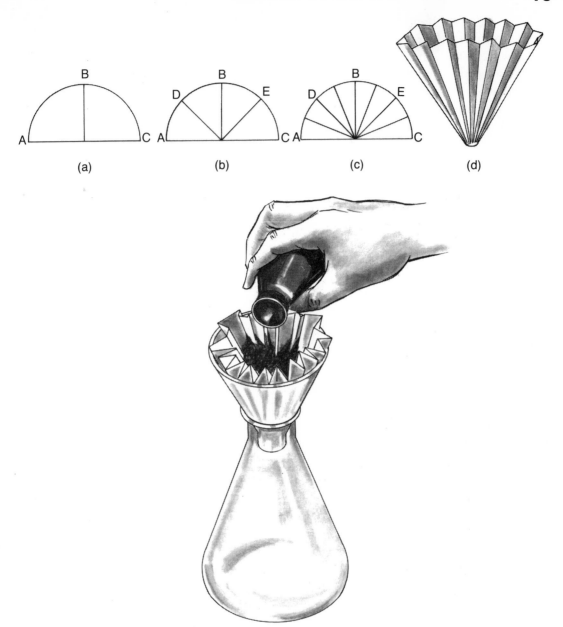

FIGURE 2.1 Preparation and use of fluted filter paper

diluted with a cosolvent and then cooled and allowed to crystallize. The size of the crystals depends on the rate of formation and the number of particles present. Shaking the solution during crystallization greatly speeds up the process and gives correspondingly smaller crystals. This is generally desirable, since solvent molecules may be entrapped and are more easily removed the smaller the particle size. In delicate selective crystallizations, however, it may be crucial to allow crystals to grow slowly without agitation.

Crystallization sometimes involves the initiation and growth of crystals from an oil or supercooled liquid by scratching or rubbing the container wall at the surface of the liquid. Crystals obtained in this way can be used as seed in a recrystallization. If crystallization of a solution cannot be readily induced, it is usually evaporated to a thin syrup containing very little solvent, but not to the consistency of a glass. Crystal growth is slower in the more viscous solution and cooling is usually not helpful. Patience is the ultimate weapon; crystallization of certain sugars has required a number of years. One of the most satisfying experiences of an organic chemist is the crystallization of a stubborn oil, and the simple fact that crystals can be obtained is often a very important point in the characterization of a compound. The formation of crystals is by no means proof that a substance is a single compound, but it is an encouraging and significant event in the isolation, for example, of a complex substance from a plant extract.

The crystals are collected by swirling and pouring the magma into a porcelain Büchner funnel (Fig. 2.2) or Hirsch funnel (small, conical shape). The funnel is fitted with a rubber stopper or conical sleeve which is seated in the neck of a heavy-wall side arm filter flask connected to an aspirator. If a small volume of liquid is to be filtered, a test tube can be placed inside the filter flask or a side arm test tube, and the stem of the funnel directed into the tube (Fig. 2.3).

These porcelain funnels are fitted with a circle of filter paper; the perforations permit liquid to drain rapidly into the evacuated flask. Be sure that the filter paper is smooth. Trim it if necessary so that it covers the per-

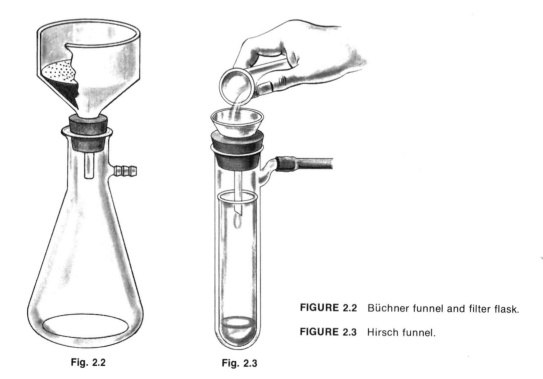

FIGURE 2.2 Büchner funnel and filter flask.

FIGURE 2.3 Hirsch funnel.

Fig. 2.2   Fig. 2.3

forations completely but does not crease into a channel at the side wall of the funnel. Wet the paper with solvent and turn on the aspirator to seat the paper firmly. After transferring the mixture to the funnel, scrape loose any solid clinging to the wall of the flask and rinse it in with a small portion of the filtrate or a little more solvent.

In small-scale crystallizations in a test tube, crystal growth may occur entirely on the walls, permitting the mother liquor to be decanted cleanly without the necessity of transferring the solid.

The next step is washing the crystals free of mother liquor. If the mother liquor is viscous, this operation is just as important as the crystallization itself. When a single solvent has been used, small portions of fresh solvent, chilled if necessary, are poured onto the crystal mass. If thorough washing to remove oil is needed, the suction should be broken and the crystals stirred well on the funnel before allowing the liquid to drain through. If a mixed solvent has been used in the crystallization, it may be desirable to use a mixture richer in the less powerful solvent for washing, but in doing this care must be taken not to precipitate impurities from the residual mother liquor onto the crystals.

The crystals should be compacted with a spatula to express as much solvent as possible. If the crystalline mass is not too dense, it may be possible to remove most of the residual wash solvent by allowing the aspirator to pull air through the funnel for a few minutes. If water must be removed, heat a watch glass slightly larger than the diameter of the funnel by passing it back and forth in a low flame. Place this over the funnel to hasten drying. (This is practical only with a small funnel; a large watch glass will invariably crack.) Transfer the crystals to glassine paper (not filter paper) for evaporation of the last traces of solvent and weighing.

wt of test tube (small = 8. 829

EXPERIMENTS

Preparation and Purification of Acetanilide

In this experiment, a simple acetylation reaction is carried out to convert an impure liquid amine, aniline, to a solid product, acetanilide, which is readily purified by crystallization.

$$NH_2 \quad + \quad (CH_3CO)_2O \quad \longrightarrow \quad NHCOCH_3 \quad + \quad CH_3CO_2H$$

aniline   acetic anhyd-   acetanilide   acetic acid
            ride (d 1.08)

The acetanilide precipitates as a crude solid which is then recrystallized from water with charcoal treatment.

Procedure. Measure ml of technical grade aniline in a 10 ml graduated cylinder. (Aniline is a toxic substance, and care should be taken to avoid contact with the skin.) Weigh the cylinder (±0.1 g), pour the aniline into a 250 ml Erlenmeyer flask and reweigh the cylinder to determine accurately the amount used in the reaction. Add 30 ml of water to the flask and then, while swirling the flask, add 5 ml of acetic anhydride in several small portions. Record any changes.

The crude acetanilide is now recrystallized in the same flask. Add 100 ml of water and a boiling stone, and heat with a burner over a wire gauze until all the solid and oil have dissolved. Remove the flame, pour a few ml of the hot solution into a small beaker and set this aside to cool. (Handle flask with a towel.) Before the boiling is resumed, add about 1 g (approximately 1/2 teaspoonful) of activated charcoal to the main solution. Do not add the charcoal to a vigorously boiling solution, or frothing will occur. Swirl the mixture and boil gently for a few minutes. Meanwhile, set up a stemless funnel or polyethylene powder funnel with fluted paper for filtration into a 250 ml or 500 ml Erlenmeyer flask. Have available about 50 ml of boiling water for washing.

Warm the funnel by pouring in a few ml of hot water; then filter the solution a little at a time. Keep the solution boiling gently until it is poured into the funnel; if crystallization begins in the funnel, add hot water to dissolve. Rinse the flask and solids in the funnel with a little hot water and set the filtrate aside to cool. To complete the crystallization, chill the flask in an ice bath for 10 to 15 minutes and prepare a Büchner funnel. Collect the crystals as previously described, dry as thoroughly as possible on the funnel, and complete the drying by spreading on paper.

In a Hirsch funnel collect the crystals that were not treated with charcoal and compare the color with that of the main batch.

Record the weight, percentage yield, and melting point of the dried acetanilide, place in a vial, label, and store for use in a later experiment (Chapt. 7).

Crystallization of n-Propyl N-Phenylcarbamate

This compound, $C_6H_5NHCO_2C_3H_7$, is a typical alcohol derivative of the type that will be encountered later in the course. The sample to be used is simply a solution of the carbamate in excess alcohol as it would normally

M.W of aniline = 93.1 g
M.w of acetanilide = 135. 2 g.
wt of aniline used = 4. 2/g.

theoretical yield
= 1 35.2/93.1 × 4. 2/ = 6. 1/g.

CHAPTER 2 – QUESTIONS **19**

be prepared. The compound is rather low melting, and the major point in this experiment is to obtain the compound in the form of crystals which can be collected and washed.

% yield = 4.82 (wt. of product) / 6.11 × 100 % = 79 %.

Procedure. Place 1 ml of the solution in a 25 × 100 mm test tube and evaporate to an oil at reduced pressure with an aspirator tube and with steam bath heating–shake constantly to avoid splashing (see page 29 and Fig. 3.7). Add several drops of petroleum ether (hexane) to the oil. If an oily layer separates, add a drop of ethyl ether to furnish a homogeneous syrup, and chill. If oil again separates, add another drop or two of ethyl ether. Crystallization should be induced in a syrup of such composition that crystals grow from a homogeneous mother liquor. This can only be achieved by arriving in trial and error fashion at the right combination of amounts of the two solvents and temperature. After crystallization has been induced, additional drops of petroleum ether can be added to decrease the solubility without causing oil to separate. Collect the crystals on a Hirsch funnel and wash with petroleum ether. Report the melting point and weight.

QUESTIONS

⋆1. Look up in a handbook of chemistry melting points and any solubility data available for the following compounds:

a. Pentaerythritol $\quad C(CH_2OH)_4$

b. *m*-Dinitrobenzene

c. Benzanilide

d. Anthracene

e. *p*-Bromoaniline

f. Camphor

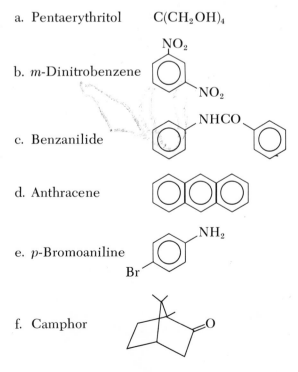

Suggest which of the solvents in the following list, or combinations of these solvents, might be suitable for recrystallization of each compound: benzene, ethanol, ether, hexane, water.

★2. Describe how you would purify the major component in the following mixture by recrystallization:

a. 90 per cent *para*-nitrobenzoic acid and 10 per cent of the *ortho* isomer.
b. 80 per cent *meta*-nitrobenzoic acid and 20 per cent of the *para*.
c. 70 per cent naphthalene ($C_{10}H_8$, mp 80°) and 30 per cent silica.

★3. During a recrystallization, an orange solution of a compound in hot alcohol was treated with charcoal and the mixture was then filtered through a fluted paper. The filtrate on cooling gave *grey* crystals, although the compound was reported to be colorless. Explain why the crystals were grey, and describe what steps would be taken to obtain a colorless product.

4. In the recrystallization of acetanilide, water is an effective solvent for the purification, permitting easy removal of insoluble impurities, but it has the drawback that the difference in solubility of acetanilide at 100° and 0° is not as large as that normally available:

Temp.	0°	80°	100°
Solubility, g/100	0.53	3.5	5.5

This means that a rather large volume of water is needed and that a significant amount of product remains in the mother liquor. From the volume of the filtrate in your experiment, calculate the amount of acetanilide remaining.

5. In the acetanilide preparation, water was added initially, and then acetic anhydride, followed by a larger volume of water. Can you see any advantage in this procedure as opposed to adding the acetic anhydride first, or all of the water first?

Reference

Crystallization

R. S. Tipson, Chapt. III. *In* A. Weissberger (Ed.), *Techniques of Organic Chemistry*, 2nd Ed., Vol. III, Part 1. Interscience, New York, 1956.

CHAPTER 3

EXTRACTION

One of the main concerns of experimental organic chemistry is the separation of mixtures and the isolation of compounds in as pure a form as needed for subsequent use. Crystallization is a useful method for purification and isolation, but it is restricted to solids, and as indicated in Chapter 2, it is a relatively inefficient way to separate a mixture of very similar compounds.

Several more general separation methods are described in this and the following four chapters. All depend in some way on the *partitioning* of the compounds to be separated *between two distinct phases.* By choosing the phases so that the different compounds are *unequally distributed* between them, fractional separation of the compounds is effected. The various methods are mechanically quite different, but they all depend on this common principle, and it should be thoroughly understood. The principle is most easily illustrated by the general method of extraction, and is therefore covered in some detail in this chapter.

THEORY OF EXTRACTION

Extraction is the general term for the recovery of a substance from a mixture by bringing it into contact with a solvent which preferentially dissolves the desired material. The initial mixture may be a solid or liquid, and various techniques and apparatus are required for different situations. In synthetic organic chemistry, the reaction product is frequently obtained as a solution or a suspension in water along with inorganic and other organic byproducts and reagents. By shaking the aqueous mixture with a water-immiscible organic solvent, the product is transferred to it and may be recovered from it by evaporation of the solvent.

The extraction of a compound from one liquid phase into another is an equilibrium process governed by the solubilities of the substance in the two solvents. The ratio of the solubilities is called the *distribution coefficient,* $K_D = C_1/C_2$, and is an equilibrium constant with a characteristic value for any compound and pair of solvents at a given temperature.

Let us consider the following situation: A 100-ml aqueous solution contains 1 gm each of compounds A and B, whose solubilities in water and ether are given below.

COMPOUND	SOLUBILITY IN WATER	SOLUBILITY IN ETHYL ETHER
A	10 g/100 ml	1 g/100 ml
B	2 g/100 ml	10 g/100 ml

By the foregoing definition the distribution coefficients of A and B (C_{ether}/C_{water}) are 0.1 and 5, respectively. If the aqueous solution is shaken with 100 ml of ether, the amount of each compound transferred to the ether phase, X, can be calculated as follows:

COMPOUND A	COMPOUND B
$$\frac{C_E}{C_W} = \frac{X_A/100}{(1 - X_A)/100} = 0.1$$	$$\frac{C_E}{C_W} = \frac{X_B/100}{(1 - X_B)/100} = 5$$
$X_A = 0.091$ g in ether	$X_B = 0.833$ g in ether
$1 - X_A = 0.909$ g in water	$1 - X_B = 0.167$ g in water

If the same aqueous solution is extracted with the same amount of ether, but in four 25 ml portions, we have for the first extraction:

$$\frac{X_A/25}{(1 - X_A)/100} = 0.1$$ $$\frac{X_B/25}{(1 - X_B)/100} = 5$$

(1) $X_A = 0.0244$ g in ether leaving 0.9756 g in water

$X_B = 0.556$ g in ether and 0.444 g in water

and for the second 25-ml extraction:

$$\frac{X_A/25}{(0.9756 - X_A)/100} = 0.1$$ $$\frac{X_B/25}{(0.444 - X_B)/100} = 5$$

(2) $X_A = 0.0238$ g in ether leaving 0.9518 g in water

$X_B = 0.247$ g in ether and 0.197 g in water

and after the third and fourth extractions:

(3) $X_A = 0.0232$ g in ether $X_B = 0.109$ g in ether

(4) $X_A = 0.0226$ g in ether $X_B = 0.049$ g in ether

Thus, the total amounts of A and B transferred to the 100 ml of ether are 0.094 g and 0.961 g, respectively. It can be seen from these values that even with a relatively small distribution coefficient ($K_D = 5$), virtually complete extraction of compound B can be effected, and that several extractions with small volumes of extractant are more efficient than a single extraction with a large volume.

If the distribution coefficient is less than about 5, a larger total volume of extractant is required for effective extraction, and if $K_D \leq 1$, the normal method of shaking the mixture and separating the layers becomes impractical. For such extractions, apparatus such as that shown in Figure 3.1 is used. The distilling solvent serves as an essentially unlimited source of fresh extractant, and the extracted compound collects in the distilling flask.

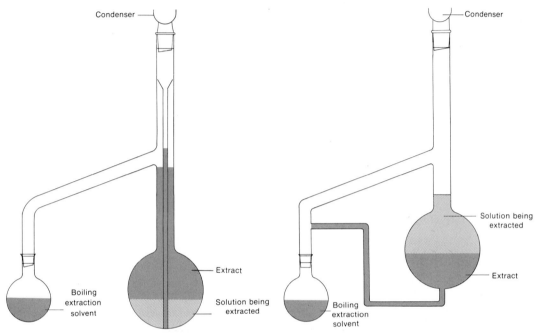

FIGURE 3.1 Continuous liquid-liquid extraction apparatus. Left, lighter-than-water solvent. Right, heavier-than-water solvent.

It may be seen in the extractions described above that a significant amount of compound A was also transferred to the ether layers. If the ether were removed from the solution at this point, the residue would be 1.055 g of material that is only 92 per cent pure B. However, if the ether is shaken (back-extracted) with 50 ml of water before evaporation, the amounts (Y) of A and B removed from the ether solution can be calculated as follows:

$$\frac{C_E}{C_W} = \frac{(0.094 - Y_A)/100}{Y_A/50} = 0.1 \qquad \frac{(0.961 - Y_B)/100}{Y_B/50} = 5$$

$Y_A = 0.078$ g in water
leaving 0.016 g in ether

$Y_B = 0.087$ g in water
and 0.874 g in ether

Evaporation of the ether now leaves a residue of 0.890 g that is > 98 per cent pure B. Some B was lost in this process, but that which remains may be pure enough for its intended use.

As mentioned, the mixture to be extracted may be a solid. Chemists commonly use this method to obtain certain naturally occurring compounds from the dried leaves, bark, or wood of various plants. Chemists and non-chemists alike brew their coffee in this way.

COUNTER-CURRENT EXTRACTION

If two compounds in a solution have nearly equal distribution coefficients, and if they must be separated by extraction, it is clear that the foregoing procedure will be very inefficient. In such instances, a technique known as *counter-current extraction* is used. To understand its principle, which is important in other applications also, consider a set of ten identical separatory funnels each containing 50 ml of water (Fig. 3.2). The mixture which is to be separated consists of a 50-ml chloroform solution of 1 g each of compounds M and N, whose distribution coefficients ($C_{chloroform}/C_{water}$) are 0.5 and 2.0, respectively.

The solution is added to the first funnel and shaken with the water. The resulting distribution of compounds is:

COMPOUND M	SOLVENT	COMPOUND N
.33 g	chloroform	.67 g
.67 g	water	.33 g

The chloroform layer from the first funnel is transferred to the second funnel, 50 ml of chloroform is added to the first, and both are shaken. The resultant distribution of compounds is summarized as follows:

COMPOUND M	FUNNEL	COMPOUND N
chloroform/water		chloroform/water
.11 g/.22 g	1	.44 g/.22 g
.22 g/.44 g	2	.22 g/.11 g

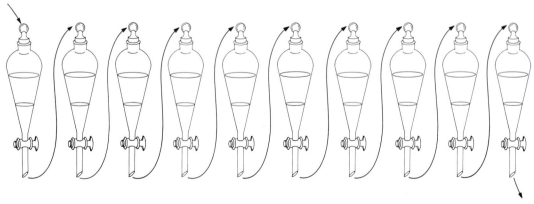

FIGURE 3.2   Counter-current extraction.

The preceding operation is repeated eight times. Each time transfer the organic layers to the next funnel, add 50 ml of fresh chloroform to the first, and shake (Fig. 3.2). At the end of this procedure, all ten funnels will contain 50 ml each of chloroform and aqueous solution. This distribution of compounds M and N among the funnels is shown in Figure 3.3. In each funnel, M is twice as concentrated in the aqueous layer as in the organic layer, and *vice versa* for N.

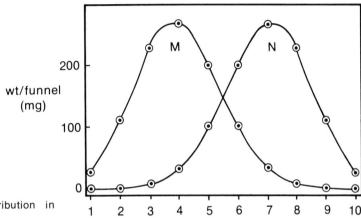

FIGURE 3.3 Initial distribution in funnels in Figure 3.2.

In order to get the compounds out of the funnels and, at the same time, effect further partitioning, the extraction is continued as follows. The chloroform layer is removed from the tenth funnel and evaporated to obtain what we shall call fraction 1. The remaining chloroform layers are sequentially transferred to the next higher funnel (9 → 10, 8 → 9 . . . . . 1 → 2); 50 ml of chloroform is added to funnel 1, and all ten are shaken as before. Fraction 2 is now removed from the tenth funnel, and the foregoing steps are repeated as often as necessary, with fresh solvent being added each time to funnel 1 and another fraction collected from funnel 10. The amounts of compounds M and N in each of the first 30 fractions are plotted in Figure 3.4. Fractions 1 to 7 contain 0.75 g of 98 per cent pure N, and fractions 14 to 30 contain 0.74 g of 98 per cent pure M.

The procedure described in Figure 3.2 would be terribly tedious to perform manually. Apparatus (Craig machine) is available, however, that automatically shakes and makes transfers between several hundred separatory tubes. This equipment provides an extremely powerful method of separation. The large number of partition stages allows the separation of compounds with very similar distribution coefficients.

In laboratory practice, most applications of the general principle of extraction involve situations in which the difference in distribution coefficients is enormous, as in the separation of an organic compound and an

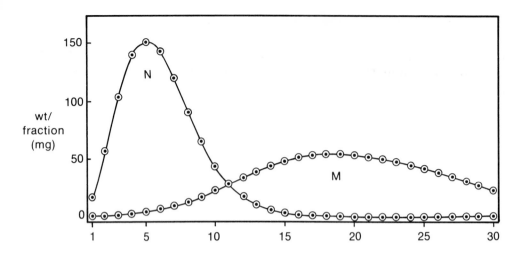

FIGURE 3.4 Fractions collected from funnels in Figure 3.2.

inorganic impurity by "washing" an organic solution with water. Another example is illustrated in the experiment in this chapter, in which the distribution coefficient is drastically altered by converting a covalent compound to an ionic species.

STEPS AND TECHNIQUE OF EXTRACTION

Use of Separatory Funnel. A separatory funnel is expensive and fragile, and when full, it is top heavy. The funnel should be supported on a ring of the proper size at a convenient height; don't prop it up on its stem. Before each use, check that the stopcock is seated and rotates freely. A clip or leash should be used to prevent the stopcock from falling out if it is accidentally loosened. A *very light* film of stopcock lubricant should be applied around a glass stopcock in bands on each side of the hole. (Teflon stopcocks require no lubricant.) Excess grease will be washed away by organic solvents and contaminate the solution.

The separatory funnel should be filled to no more than about three-fourths of the total depth, so that thorough mixing is possible. After filling (check first that stopcock is closed!), stopper securely with a properly fitting plastic, glass, or rubber stopper. (Before using the funnel for the first time, it is a good idea to shake with a few milliliters of liquid to make sure that stopcock and stopper are tight.)

Hold the funnel with the stopcock end tilted up; the stopper is kept in place securely with the heel of one hand, and the stopcock end is supported in the other hand (Fig. 3.5). As soon as the funnel is inverted, open the stopcock to release any pressure. Then close the stopcock and shake in a horizontal position for about 30 seconds. Stop and slowly open the stopcock a few times to vent any pressure that may have built up; this is particularly important in extractions with bicarbonate, when CO_2 pressure may develop during the extraction. Replace the funnel in the ring and loosen the stopper. When the phases have completely separated, draw off the lower layer through the stopcock.

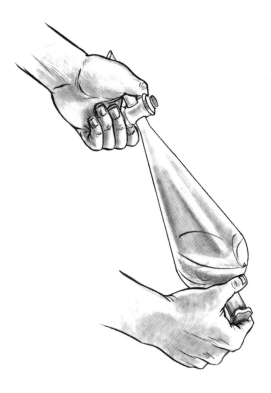

FIGURE 3.5 Use of separatory funnel.

In the extraction of an aqueous solution, the solvent may be either lighter (e.g., ether) or heavier (e.g., chloroform) than water. In the first case, if several portions of solvent are used, the aqueous layer must be drained into a receiver (usually the flask in which it was originally contained), and the ether solution transferred to a second flask. The aqueous phase is then returned to the separatory funnel for further extraction as needed. With a solvent denser than water, the aqueous solution is simply retained in the funnel and shaken with successive portions of solvent. In either case, the organic layers are usually then combined, returned to the funnel, and shaken with a small volume of water to remove traces of the original aqueous phase that are suspended in the organic layer.

It is important to note that an aqueous solution to be extracted often contains a water-miscible solvent such as alcohol. If the amount of alcohol is significant, an excessive volume of extracting solvent (such as ether) is required to form two liquid phases, and the ether layer will contain a substantial fraction of the third solvent as well as water. If this situation arises, it may be possible to evaporate some of the alcohol before extracting. Otherwise, a larger-than-normal amount of ether must be used in the extraction, and the ether phase is then thoroughly back-extracted with water.

In many cases the separatory funnel is used simply as a means of recovering an organic product from a large amount of water with minimum mechanical loss. A small volume of ether or other solvent is added to the mixture to permit sharp separation of layers. Even though the compound may have a negligible solubility in water, a second portion of solvent should be used to rinse the aqueous layer and the separatory funnel.

A common application of extraction is the removal of water-soluble impurities from an organic solution. For example, an ether solution may contain dissolved hydrogen chloride. This is removed by "washing" the ether in a separatory funnel with aqueous carbonate or hydroxide solution and then water.

Drying Agents.   After any of these extraction processess, the organic solution is saturated with water, and it is desirable to dry the solution before evaporating the solvent. Water is an impurity and should always be removed before a crystallization procedure is carried out or before a liquid is distilled. A number of salts that form hydrates can be used for this purpose. The efficiency of a drying agent depends on the completeness of drying (intensity), the degree of hydration (capacity), and the rate at which the salt absorbs water. A few of the more commonly used agents are listed here.

1. Magnesium sulfate has high capacity and intermediate intensity. It is cheap and rapid in action and is the most generally useful all-purpose drying agent.

2. Sodium sulfate has high capacity but low intensity; it can be used only at low temperature.

3. Calcium chloride has high intensity, but is useful primarily for hydrocarbons or halides because of complex formation with most compounds containing oxygen or nitrogen.

4. Calcium sulfate (Drierite) has very high intensity and is rapid, but it has low capacity.

5. Potassium carbonate is often used for drying basic compounds; it has intermediate intensity and capacity.

6. Molecular sieves are complex silicates with a porous structure that selectively entraps water molecules. They are useful for obtaining rigorously anhydrous liquids.

For solutions, the amount of drying agent should usually be about one-tenth of the liquid volume; a smaller amount is used for a pure liquid. After allowing the solution to stand over the drying agent, swirling occasionally, for 10 to 20 minutes, remove the drying agent by filtering through a cotton plug and rinsing with a solvent.

Evaporation of Solvent.   An important operation in extractions and other procedures such as column chromatography (Chapt. 7) is the evaporation of a solvent to permit recovery of a relatively nonvolatile residue. Ether, methylene chloride, or hydrocarbon solvents are flammable and toxic. If more than a few ml of solvent is to be removed, provision must be made to avoid the discharge of vapor into the room. A convenient way to accomplish this is to sweep the vapor, as it distills, into the aspirator. This can be done by inserting a piece of glass tubing into the rubber tubing leading to the aspirator and clamping it vertically above the flask on the steam bath, with the glass tubing extending a short distance into the neck (Fig. 3.6). The pressure in the system is not reduced, but most of the vapor will enter the aspirator where it is diluted with water and discharged into the drain. This is inadequate for large volumes of solvent, which must be removed by distillation through a condenser.

For rapid evaporation with minimum heating, the solution is placed under reduced pressure in a round-bottom flask or test tube. This is very easily arranged by fitting a large one-hole rubber stopper over the glass tubing on the aspirator tube and placing the stopper on the mouth of the flask (Fig. 3.7). Atmospheric pressure holds the stopper in place, but it can be quickly released. When this method is used, the flask or test tube must be no more than 10 to 15 per cent full, and it must be agitated constantly, with the pressure controlled by placing a thumb on the tubing to prevent bumping and frothing. This technique requires a little practice, but it will be found to be a great time saver and well worth learning.

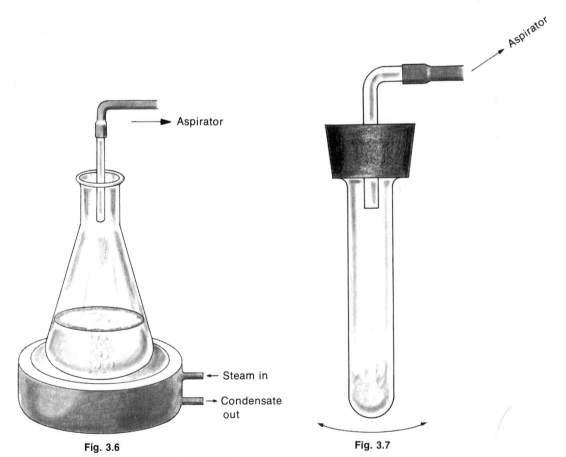

Fig. 3.6

Fig. 3.7

FIGURE 3.6 Evaporation of solvent on steam bath.

FIGURE 3.7 Evaporation of solvent at reduced pressure.

For rapid evaporation of larger volumes of solvent, rotary evaporators are standard equipment in most advanced laboratories. These evaporators are operated at reduced pressure, with a very efficient condenser. The flask is held at an angle and rotated rapidly in the heating bath; this action spreads the liquid in a film on the inside wall and provides a large surface for evaporation.

The correct size and type of glassware for evaporation deserves comment. There must be room for boiling and agitation, whether under reduced pressure or not, and the flask should never be more than one-fourth to one-third full. Much time can be wasted by gingerly evaporating a solution from a flask that is too full. Evaporation of solvent should never be done in a beaker. Although it has a wide mouth, a beaker is actually a very inconvenient vessel from which to remove a crystalline residue or a liquid, since solvent tends to creep up the walls, and rinsing is less efficient; moreover, a beaker cannot be stoppered, swirled, or evacuated. Vessels containing liquids to be dried or stored should always be securely stoppered with corks to prevent evaporation and contamination. A beaker is really useful only as a container for weighing out a solid or as a bath for cooling or heating. For practically all other purposes, Erlenmeyer flasks, round-bottom flasks, or test tubes are preferable.

Safety Note: Only a *spherical* or *cylindrical* vessel (e.g., test tube) should be placed under reduced pressure unless it is made of heavy wall glass, such as a filter flask. A flat-bottom surface is not built to withstand the force exerted by nearly 1 atmosphere pressure if the flask is evacuated. In a round-bottomed flask or test tube, the pressure is distributed over a continuously curved surface. Most modern glassware is being made with fairly thick walls, but it is *extremely* hazardous to evacuate a large Erlenmeyer flask because of the danger of implosion and flying glass.

It is frequently necessary to weigh a residue after evaporating a solvent or decanting a liquid from a solid; the empty weight, or *tare*, of the vessel should be determined in advance. It is a good idea to identify by some permanent marking (e.g., a small number scratched with a diamond pencil or carborundum stylus) a number of 25 × 100 mm test tubes, and to record the tare weight for future reference. This tare is invalidated, of course, if the tube becomes chipped or heavily scratched.

Transfer of Liquids. A few points of technique are important in manipulating small volumes of organic liquids and solutions. It is usually necessary to transfer a solution to a smaller vessel after evaporation is about 80 per cent complete to facilitate removing the last trace of solvent and to minimize losses in recovering a crystalline residue.

A small volume of an organic liquid should be transferred with a pipet (p. 6). An attempt to pour it will result in a significant fraction of the residual compound being spread over the inside wall, the lip, and the outer surface of the neck because of the low surface tension of the organic solvent. An excessive amount of solvent is then required for rinsing. In rinsing a relatively large flask with ether or methylene chloride, add a little solvent and then warm the bottom of the flask so that the condensing vapor rinses down the walls; then concentrate the rinse in a small pool for transfer.

EXPERIMENT   *OH on big benzene is acider.*

Separation of *p*-Dimethoxybenzene and *p*-Methoxyphenol

This experiment illustrates the use of extraction in a very common situation—the separation of neutral and acidic compounds arising from a reaction mixture. In the conversion of hydroquinone to the monomethyl ether with dimethyl sulfate, it is difficult to avoid the formation of the dimethoxy compound, and the latter must be separated before the desired product is isolated.

Hydroquinone	*p*-Methoxyphenol	*p*-Dimethoxybenzene
	mp 55–56°	mp 56°

The separation depends on the fact that phenols are weak acids and in an aqueous base are converted to the phenoxides which are soluble in water. By shaking an ether solution of a phenol in a separatory funnel containing aqueous NaOH, the phenol is converted to the phenoxide and is extracted from the organic phase. After separation of the phases, the phenol is recovered from the aqueous solution by acidification, and extraction is again employed to remove it from the aqueous solution.

Procedure. Transfer about 0.6 g of the methoxyphenol-dimethoxybenzene mixture to a tared 25 × 100 mm test tube and weigh to ±0.01 g. Dissolve the sample in about 15 ml of ethyl ether and pour into a separatory funnel. Rinse the test tube with several small portions of ether and then add more ether to make a total volume of about 50 ml.

Shake the ether solution successively with two 10-ml portions of 0.5N NaOH and then with 10 ml of water, combining these three aqueous extracts. Label as "phenol fraction" and set aside.

Pour the ether solution out of the top of the separatory funnel into a 125 ml Erlenmeyer flask. Rinse the funnel with a few milliliters of ether. (*Note:* Although it might seem more sensible to drain the ether through the stopcock, water entrapped in the stopcock bore and clinging to the wall will also drain out.) Add a spoonful of $MgSO_4$ to this solution, label as "neutral fraction," stopper, and set aside for 15 minutes; swirl occasionally.

Pour the aqueous "phenol fraction" back into the separatory funnel, rinsing the flask with a few ml of water. To this solution add 2 ml of 6N hydrochloric acid. Check the pH of a drop of this solution with test paper; if the solution is not acidic, add a little more acid to produce an acidic pH. Extract with three 10-ml portions of ether. (Review the detailed description of extraction with a lighter-than-water solvent on p. 27.) Wash the combined solutions with 3 to 4 ml of water, transfer this ether solution to an Erlenmeyer flask, as described above, add drying agent, and label as "phenol."

Filter the drying agent from the "neutral fraction" through a loose cotton plug into a 250 ml round-bottom flask; evaporate this solution to small volume using reduced pressure (p. 29). Transfer to a tared test tube, evaporate to dryness (Fig. 3.7), and crystallize. Record the weight and melting point of the "neutral fraction."

Isolate the phenol in the same way, determine the melting point, and report the total recovery of both compounds and the per cent of dimethoxybenzene in the mixture.

QUESTIONS

★1. In the back-extraction described on page 23, calculate the amount of A and B remaining in the ether layer after using two 25-ml portions of water instead of one 50-ml wash.

★2. In the experiment in Chapter 2, the crude precipitate of acetanilide was recrystallized directly from the aqueous mixture. This procedure depended on the favorable solubility of acetanilide in water and is not generally applicable; in most instances, the crude product would be isolated by extraction. Describe, with details, a procedure for isolating the acetanilide based on extraction.

★3. In an oxidation experiment, 1 g of a neutral, water-insoluble, ether-soluble ketone was obtained along with some acidic byproducts in 10 ml of acetone solution. The ketone was recovered by extraction. Describe all steps in the isolation procedure.

4. Carboxylic acids such as benzoic acid are *stronger acids* (more extensively dissociated) than phenols. Carbonic acid has an intermediate acidity.

Benzoic acid: $C_6H_5CO_2H + H_2O \rightleftarrows C_6H_5CO_2^- + H_3O^+$
$$K_a = 6 \times 10^{-5}$$

Phenol: $C_6H_5OH + H_2O \rightleftarrows C_6H_5O^- + H_3O^+$
$$K_a = 1 \times 10^{-10}$$

Carbonic acid: $H_2CO_3 + H_2O \rightleftarrows HCO_3^- + H_3O^+$
$$K_a = 3 \times 10^{-7}$$

From these data, it is seen that benzoic acid is soluble in bicarbonate solution:

$$C_6H_5CO_2H + HCO_3^- \rightarrow C_6H_5CO_2^- + H_2O + CO_2$$

Devise a procedure for the separation of *p*-chlorobenzoic acid, *p*-chlorophenol, and *p*-dichlorobenzene based on an extraction process. (All three compounds have negligible solubility in water.)

5. Both *p*-methoxyphenol and *p*-dimethoxybenzene are solids at room temperature. What was the physical state of the phenol when it was liberated from the phenoxide by addition of HCl? Explain.

Reference

Extraction

L. C. Craig, and D. Craig, Chapt. 2. *In* A. Weissberger (Ed.), *Techniques of Organic Chemistry,* 2nd Ed., Vol. III, Part I. Interscience, New York, 1956.

4

DISTILLATION

Distillation is the process of vaporizing a substance, condensing the vapor, and collecting the condensate in another vessel. In this way, mixtures of liquids with different volatilities can be separated or recovered from nonvolatile contaminants. When vaporization occurs from a solid, the process is called *sublimation* (see Chapt. 19). The term *evaporation* is used in laboratory practice when the residue, rather than the distillate, is of importance and the distilling vapors are permitted to escape uncondensed into a hood or an aspirator. The term *reflux* means vaporization with return of the condensate to the original flask.

The basic elements of a distillation apparatus are a boiling flask, a column through which the vapor rises, a thermometer, a condenser, and a receiver. The liquid is heated until its vapor pressure equals that of the atmosphere and boiling commences. In the distillation of a pure liquid, vapor from the boiling flask rises and condenses until the apparatus is in thermal equilibrium and the thermometer registers the boiling point, which will remain constant as long as liquid and vapor are both present.

The situation that prevails in the distillation of a mixture of two miscible liquids can be seen in a vapor-liquid phase diagram such as Figure 4.1. The boiling points of pure compounds A and B are T_A and T_B. The lower curve gives the boiling point of various mixtures of A and B, and the upper curve, the condensation point of mixtures of gaseous A and B. These two lines do not coincide at a given temperature (and pressure), and the compositions of the liquid and vapor phases in equilibrium differ. Specifically, if an equimolar mixture of liquids A and B (composition X_1) is heated to its boiling point (temperature T_1), the vapors leaving the liquid will have a composition (X_2) which is enriched in the lower-boiling component (approximately 80 per cent A in Fig. 4.1). If a sample of this vapor is removed, the composition of the remaining liquid shifts to the right on the diagram; i.e., the relative amount of B increases, and the mixture has a higher boiling point. Continued distillation from the mixture provides distillate progressively enriched in B, but each fraction of it will contain both components. The net result is that a *simple distillation* of two liquids, whose boiling points differ by less than about 50°, effects relatively little

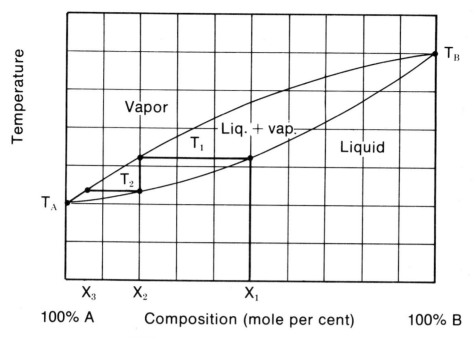

FIGURE 4.1 Vapor-liquid phase diagram.

separation, and the temperature will rise continuously during the distillation (dashed line in Fig. 4.2).

Further study of Figure 4.1 will illustrate how the separation efficiency of a distillation can be increased. If the condensed vapors of composition X_2 are reequilibrated at temperature T_2, the vapors will now have a composition X_3 which is even more enriched in the lower-boiling component A. If the equilibration can be repeated several times, it is possible to obtain a final distillate which is nearly pure A. This process, termed *fractional distillation,* is accomplished by distilling through a column with a large surface area at which repeated exchange of molecules between liquid and vapor phases occurs. A large area can be obtained in a small volume by packing the column with inert metal sponge or glass or metal helices. One of the most efficient methods is provided by a spinning-band column, in which a spiral of metallic mesh is rotated at high speed inside the column during the distillation.

A fractional distillation, carried out slowly with a good fractionating column, will result in a distillation plot resembling the solid line in Figure 4.2. The fractions collected during the level regions will be reasonably pure samples of the components of the original mixture. Thus distillation is another example of the principle whereby an inefficient separation step is repeated many times to obtain compounds of a high degree of purity.

The phase diagram in Figure 4.1 is typical of mixtures of similar compounds which form ideal or nearly ideal solutions as defined by Raoult's Law:

$$P_{total} = P_A + P_B = P_A^\circ X_A^\circ + P_B^\circ X_B^\circ$$

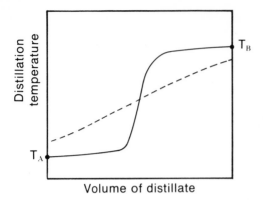

FIGURE 4.2 Distillation curves.

$P°$ and X are the vapor pressure and mole fraction, respectively, of the individual compounds. In a distillation at atmospheric pressure, $P_{total} =$ 760 mm Hg.

With mixtures of dissimilar compounds, such as an alcohol and a hydrocarbon, association of one component occurs in the liquid, and the solution usually does not follow this relationship. At some composition, in this case, there will be a minimum in the vapor-liquid phase diagram (Fig. 4.3). Such a system is called an *azeotrope,* and a liquid of composition X will distill completely at temperature T_X as a constant boiling mixture. Fractional distillation of a mixture of A and B of any other composition will provide distillate of composition X until only one component remains in the liquid. A common example of this type of azeotrope is ethanol-water (95:5).

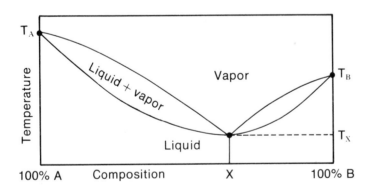

FIGURE 4.3 Phase diagram for minimum-boiling azeotrope.

APPARATUS AND TECHNIQUE

Standard assemblies for distillation are illustrated in Figures 4.4 and 4.5. The distilling flask is always supported by a ring and wire gauze. The flask with liquid (and boiling stone!) is loosely clamped and is held in place by the column, which should be perfectly vertical and securely

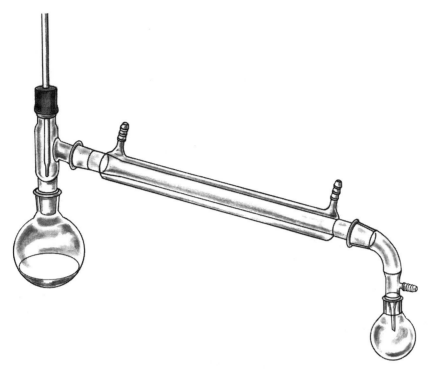

FIGURE 4.4 Simple distillation set-up.

clamped. It is undesirable to clamp securely both the flask and the column (or distilling head) since the column may tend to shift slightly when the condenser is attached, preventing a snug fit of the bottom joint and permitting hot vapor to escape just above the burner. The condenser is clamped on a second ring stand with height and angle adjusted so that the joint fits *without stress or binding* into the side arm joint. Connect the rubber tubing for condenser water with inlet at the bottom and outlet at the top. Attach an adapter and suitable receiver (Erlenmeyer or round-bottom flask or vial—never a beaker), supported by a third ring stand if necessary—do not prop receiver on a make-shift support.

In assembling ground-joint apparatus, a *very thin* film of stopcock grease may be used to ensure a snug fit. It is much more important to be sure that the joint is clean and free of grit and is properly aligned. A common error is to use too much grease; this results in contamination of any liquid that comes in contact with the joint.

It is important to select the proper size flask for a distillation. As a general rule it should be no more than two-thirds full initially. Since there is a fixed amount of "hold-up" of condensing vapor on the walls, it is undesirable to use too large a flask. The *boiling stone* is essential to maintain constant ebullition (formation of bubbles and uniform turbulence). For this purpose a small lump of carborundum or other porous mineral is used. Without this stone to provide a steady stream of gas bubbles, super-heating and bumping of the liquid will usually occur. This applies with even greater force when evaporating a liquid on a steam bath without a condenser. If the boiling stone is forgotten until the liquid is hot, and possibly superheated, the flask must be cooled below the boiling point before the stone is added or the liquid may erupt. Since the pores fill with liquid as soon as boiling ceases, a stone cannot be reused, and a fresh one must be added if the distillation is interrupted.

Simple Distillation. This type of distillation is carried out with apparatus set up as in Figure 4.4. Soon after boiling begins, a ring of condensing vapor will be seen rising in the flask and neck. The thermometer, which should be positioned as shown with the bulb just below the side arm, will then begin to register an increase until steady distillation is underway. The thermometer should be bathed in vapor with a constant drop of condensate on the tip. In this condition, it will register the temperature at which liquid and vapor are in equilibrium, i.e., the boiling point. The temperature in the distilling head will not level off to the boiling point until thermal equilibrium with the glass walls is established. This requires a minute or two, and at least a few drops of distillate will be collected before the boiling point is registered. The heating rate should be adjusted to maintain a distillation rate of about one drop per second. If heating is too rapid, an appreciable amount of distillate may appear to be lower boiling "forerun." If distillation is too slow, particularly with a high boiling liquid, the true boiling point may never be observed because of heat losses.

Safety Note: Distillation must *always* be stopped before the flask becomes completely dry. Without the absorption of heat due to vaporization, the flask temperature can rise very rapidly. Many liquids, particularly alkenes and ethers, may contain peroxides which become concentrated in highly explosive residues.

Fractional Distillation. Fractional distillation, in which a packed column is used (Fig. 4.5), is carried out in the same way. The rate of distillation should be adjusted so that the packing is wet with liquid throughout its length, but not flooded with pools of condensate. For highly efficient fractionation, the column is heated by a resistance winding or is vacuum-jacketed to lessen heat loss to the air. If these are unavailable, a loose wrapping of aluminum foil around the column is sometimes helpful. Another important refinement involves the use of a distilling head constructed so that a large fraction of the vapor reaching the head returns to the column, with only a small fraction being removed as distillate, so as to minimize disruption of the equilibrium.

Vacuum Distillation. Many organic liquids cannot be distilled at atmospheric pressure because the temperature required causes decomposition. This frequently is the case with compounds boiling much above 200° and sometimes at even lower temperatures. This difficulty can be overcome by distilling at a lower pressure. The vapor pressure of any substance is a function of temperature, and the lower the pressure within the distillation apparatus, the lower the boiling point. A plot of vapor pressure *vs* temperature has the form shown in Figure 4.6. For simple organic com-

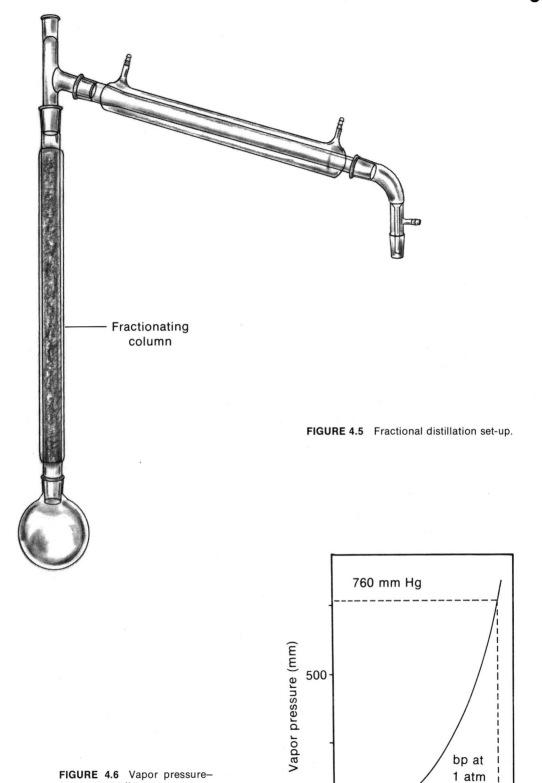

FIGURE 4.5 Fractional distillation set-up.

Fractionating
column

760 mm Hg

Vapor pressure (mm)

500

0

bp at
1 atm

Temperature

FIGURE 4.6 Vapor pressure–
temperature diagram.

pounds, reduction of the pressure from 750 to 20 mm causes a drop in boiling point of 90 to 120°. A nomogram for use in estimating boiling points of relatively nonpolar compounds at reduced pressures is given in Figure 4.7. This chart is less accurate for compounds which are highly associated in the liquid phase.

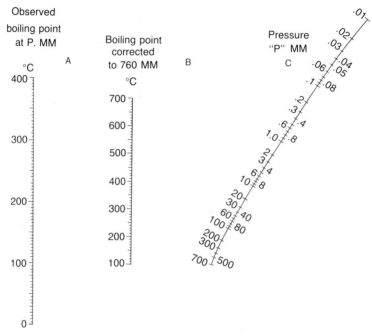

FIGURE 4.7 Boiling point–pressure nomogram. To find the atmospheric pressure boiling point of a compound for which the boiling point is known at a reduced pressure (*e.g.*, 70° at 1 mm), lay a straight edge from the known boiling point (70°) on scale A to the pressure (1 mm) on scale C. The intercept on scale B gives the boiling point at 760 mm (~240°). By connecting the atmospheric pressure boiling point on scale B (240°) with a different reduced pressure on scale C (*e.g.*, 100 mm), the boiling point at this latter pressure can be read on scale A (165°).

Distillation at reduced pressure is carried out in a set-up such as that shown in Figure 4.8. Pressures down to about 20 mm can be obtained with a good water aspirator; for lower pressure, an oil-sealed mechanical pump is used, and a cold trap must be placed in the system to protect the pump from vapors. To obtain a meaningful boiling point, a manometer must be included to measure the pressure. A Bunsen valve from a burner can be included to permit a controlled leak for distillation at an intermediate pressure. If several fractions are to be collected, some means of changing receivers without interrupting the distillation is required; this is usually a device (called a cow) which can be used to position different receivers under the outlet from the condenser (Fig. 4.8).

In vacuum distillation, it is essential to maintain steady boiling; this can be done by allowing a slow stream of air or nitrogen to leak through a capillary under the surface of the liquid or by use of a magnetic stirrer. After assembling the apparatus, the pressure is reduced with the aspirator or pump before heating is begun. The distillation is then carried out essentially as described previously.

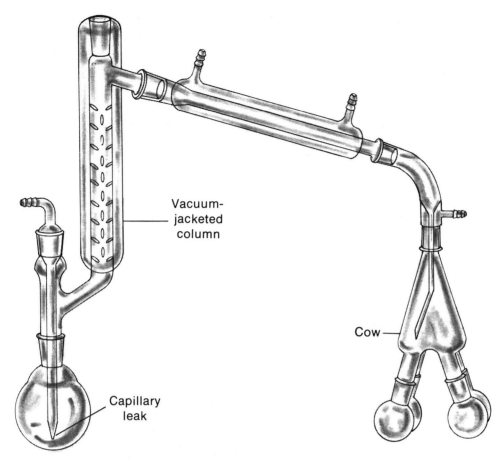

Vacuum-
jacketed
column

Cow

Capillary
leak

FIGURE 4.8   Vacuum distillation set-up.

EXPERIMENTS

1. Simple Distillation (Fig. 4.4)

Assemble a set-up without a fractionating column with a 50 ml flask containing 30 ml of a benzene-toluene mixture. Heat with a microburner or heating mantle, and distill about 25 ml of the mixture into a 100 ml graduated cylinder at a rate of about 2 ml per minute. Record the initial boiling point and the boiling point after each 2 ml of distillate is obtained.

2. Fractional Distillation (Fig. 4.5)

Reassemble set-up, this time with the fractionating column, and return the distillate from the first experiment to the flask.

Distill again, using the column, and record boiling point *vs* distillate volume as before. Collect several fractions, changing as you think appropriate to obtain the maximum amount of the two components in fractions with 2° boiling range. These fractions should be saved if the next experiment (vapor phase chromatography) is to be done.

Plot the boiling point *vs* distillate volume for both distillations on the same graph; include this in your report, together with the volumes and boiling points of the fractions obtained in the second distillation.

3. Unknown

Obtain a sample of an unknown from your instructor and distill in a fractionating column as indicated; plot the boiling point *vs* distillate volume and report results.

QUESTIONS

⋆1. If a fractional distillation of a mixture of A and B (Fig. 4.1) is carried out with a column in which there are four equilibration stages (termed *plates*), what will be the composition (±1%) of the distillate when the mixture in the distilling flask is 50 per cent B; 95 per cent B; 99 per cent B?

⋆2. When a vacuum distillation is carried out using an aspirator, what effect will fluctuations in the water pressure have on the course of the distillation?

⋆3. In contrast to the system illustrated in Figure 4.3, some azeotropes boil higher than either of the components. The boiling points of acetone and chloroform are 56.5° and 61.2°, respectively. A 1:4 mixture of these compounds has a constant boiling point at 64.7°. Sketch a phase diagram which indicates these data. In a fractional distillation of a 50:50 mixture of these compounds, what would be the boiling point and composition of the initial and final fractions of distillate?

Reference

Distillation

A. Weissberger (Ed.), *Techniques of Organic Chemistry*, 2nd Ed., Vol. IV. Interscience, New York, 1965.

CHAPTER 5

VAPOR PHASE CHROMATOGRAPHY

Vapor phase chromatography (*VPC* or *GLPC* for gas-liquid partition chromatography or simple *GC*) is a separation method which combines elements of both distillation and extraction. The term chromatography was coined for an earlier technique used to separate colored compounds (pigments) obtained in plant extracts (see Chapt. 6). VPC is a recent development compared to the other partition methods, but because of its simplicity and high efficiency, it has become very popular since its discovery in 1952.

All types of chromatography are based on the partitioning of compounds between a stationary phase (solid or liquid) and a moving phase (liquid or gas). They differ from the Craig distribution technique described in Chapter 3 in that the motion of the moving phase is continuous rather than in discrete steps. The theoretical analysis in terms of distribution coefficients is slightly different also, but the result of such a process is practically identical to that shown in Figures 3.3 and 3.4.

APPARATUS

The major components of the apparatus required for vapor phase chromatography are shown schematically in Figure 5.1.

The gas source provides the moving phase, usually nitrogen or helium, at a constant pressure at one end of the tubing containing the stationary phase. Pressures of the order of 5 to 50 psi are normal. Also at the inlet end of the tubing is an opening for introducing the mixture to be separated. This is commonly a hole sealed with a soft rubber plug or septum through which the needle of a syringe containing the sample may be inserted. After injection, the sample is vaporized by the high temperature of the area, and the carrier gas transports the vapors to the part of the column containing the stationary phase. Alternatively, and somewhat better, the injection port may be so designed that the sample is introduced directly into the end of the stationary phase.

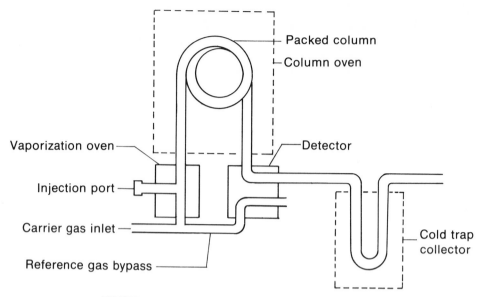

FIGURE 5.1 Schematic diagram for gas chromatograph.

The stationary phase for the partition is contained in a length of metal or glass tubing which is called a column even though it may be bent or coiled to conserve space. The tubing usually has a 2 to 4 mm inside diameter and is 5 to 50 feet long. If the compounds emerging from the column are to be collected, the tubing diameter is usually 6 to 20 mm to provide a greater capacity. The stationary phase itself consists of a liquid which is coated on uniformly sized particles of a solid to obtain a large surface area. The compounds in the sample are partitioned between the stationary liquid phase and the moving gas phase and, due to differing distribution coefficients, are transported at different rates through the column (Fig. 5.2). The distribution coefficient is obviously related to the vapor pressure of the compound and its solubility in the liquid phase. Because of the former factor, it is usually necessary to heat the column, and for this purpose it is enclosed in a thermostated oven. Since the liquid stationary phase must remain in the column, it must have a very low vapor pressure at the operating temperature. Liquids commonly used are high boiling hydrocarbons, silicone oils, and polymeric esters, ethers, and amides.

Since the distribution coefficient is a function of the compound's affinity for the stationary phase, as well as its vapor pressure, two compounds with identical boiling points may be separated if they have different solubilities in the liquid phase. Thus, n-hexane and ethoxyamine $(C_2H_5ONH_2)$, both of which boil at 68°, would be well separated by a VPC column containing poly(ethylene glycol succinate), which is a relatively polar liquid phase.

At the end of the VPC column is a device for detecting the compounds as they are eluted. The two most common types of detectors are based on thermal conductivity and flame ionization. The former measures the

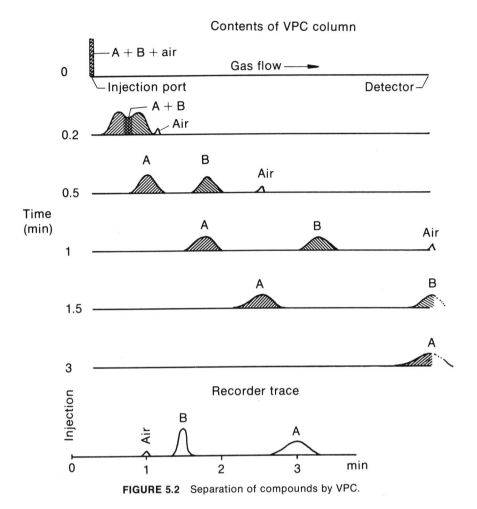

Contents of VPC column

FIGURE 5.2 Separation of compounds by VPC.

amount of cooling of a heated thermistor (an element whose electrical resistance varies with temperature) by the effluent gases. Helium is used as a carrier gas because of its very high thermal conductivity. Since the vapor of any organic compound has a lower thermal conductivity, a mixture of helium and the compound will cool a thermistor less than an equal flow of pure helium. A flame ionization detector measures, by electrical conductivity, the number of ions produced when a compound that is eluted burns in a hydrogen flame. It is much more sensitive than a thermal conductivity detector, and nitrogen can be used as the carrier gas, rather than the more expensive helium. Both of these methods measure differences in conductivity, and in each case a reference detector is included through which pure carrier gas flows. An electrical signal from the detector is transmitted to the pen drive of a strip-chart recorder.

If a compound leaving the column is to be collected, the effluent from the thermal conductivity detector can be passed through a cold trap where the vapors of the compound condense from the helium gas. Normally,

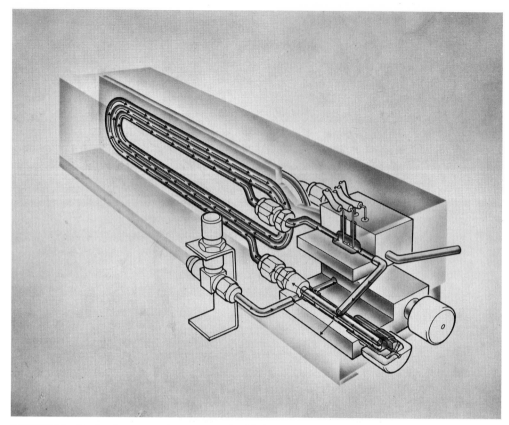

FIGURE 5.3 Dual column gas chromatograph with thermal conductivity detectors (Carle Model 8000).

VPC is used as an analytical tool, and the vapors escape into the air. A cutaway view of a simple gas chromatograph is shown in Figure 5.3.

THEORY

Despite the general similarity of VPC and chromatography to the Craig distribution discussed in Chapter 3, there are a few additional terms and concepts which are generally restricted to chromatographic separations. One of the most basic of these is the rate of transfer of a compound through the column relative to the velocity of the moving phase. This is termed R_F and is most easily visualized as the fraction of the total column length traversed by the compound in the time it takes for a unit volume of the carrier gas to flow from one end to the other.

For compounds A and B in Figure 5.2, the R_F values are .33 and .67. In this figure, and in practice, a small amount of air coinjected with the sample serves to mark the passage of the eluting gas through the detector. (*Note:* Flame ionization detectors are insensitive to air, but any combustible gas could be used for this purpose if necessary.) A more common term in VPC is the retention time, R_T, which is the time it takes from injection to elute a compound from the column. These two quantities are simply related by the formula $R_T = R°_T/R_F$, where $R°_T$ is the retention time of the air. In both of these definitions, the air is assumed to move at the carrier gas velocity, i.e., it has a distribution coefficient ($K_d = C_{liquid}/C_{vapor}$) of 0. The distribution coefficients of other compounds are given by

$$K_d = \left(\frac{1}{R_F} - 1\right)\frac{V_G}{V_L} = \left(\frac{R_T}{R°_T} - 1\right)\frac{V_G}{V_L}$$

where V_G = volume of gas in the column, and V_L = volume of liquid phase in the column, e.g., $K_d(A) = 2$ and $K_d(B) = 0.5$, if $V_G/V_L = 1$. Generally, this latter ratio is much less than one and is not readily determined in chromatography.

Another common quantity in VPC is the "height equivalent to a theoretical plate" (HETP). The term, *theoretical plate*, is derived from a type of fractional distillation column in which each simple distillation step (Fig. 4.1) is physically separated by a perforated horizontal plate. It is also analogous to each of the ten separatory funnels in the Craig apparatus (Fig. 3.2). In VPC, one HETP is the length of column in which the equivalent of one simple equilibration step occurs. Thus, the length of the column divided by the HETP is the effective number of partitions a compound undergoes in its passage through the column. The high efficiency of VPC in separating compounds is due to the fact that HETP values of one millimeter or less are possible, and a ten-foot long column may contain up to 5000 to 6000 theoretical plates. A simple formula for the calculation of the number of theoretical plates (n) in a column is

$$n = 16(d/w)^2$$

where d and w are defined in Figure 5.4 and are measured in the same units (minutes, millimeters, inches). The HETP of a given column is

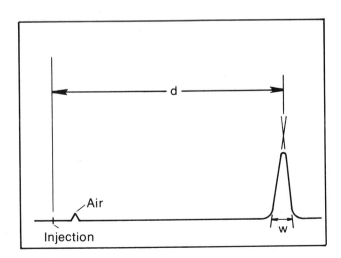

FIGURE 5.4 Calculation of number of theoretical plates.

somewhat dependent on the operating conditions (temperature, flow rate, sample size) as well as the specific compound used for the measurement. For mixtures of very similar compounds, when n is constant, the preceding equation indicates that the peak width is directly proportional to the retention time (see Chapt. 24).

TECHNIQUE

Optimum separation (resolution) of compounds on VPC is obtained with small sample sizes. For columns made with 1/8-inch tubing (0.09 in. i.d.), this means 1 to 5 microliters. For preparative work, larger amounts of samples can be handled on columns of larger diameter; the sample size can be increased in proportion to the cross-sectional area of the tubing. Thus with 3/4-inch tubing (0.68 in i.d.) up to 1 ml may be injected.

Column diameter also determines the optimum flow rate of the carrier gas; 10 to 30 ml/min is typical with 1/8-inch columns. The column temperature is usually set at or slightly below the average boiling point of the mixture. If several compounds of widely differing boiling points are to be separated, provision is usually made for increasing the oven temperature during the chromatography. This can be done stepwise or continuously.

In determining the optimum flow rate and temperature for a separation, adjustment of one or the other or both is made so that the average R_F value is between 0.1 and 0.5, and the R_T of the slowest moving compound is on the order of 5 to 10 minutes.

The sample to be separated by VPC is taken up (neat or in a low-boiling solvent) in a syringe which is calibrated in microliters. A small amount (1 to 2 μl) of air should be drawn in after the liquid so that the needle is empty. This is done to avoid any vaporization of the sample from the needle prior to injection of the remainder in the barrel of the syringe. The needle of the syringe is inserted through the septum into the injection port, and the sample is injected, taking care not to bend the fragile plunger in the process. The injection should be rapid in order to get all of the sample started through the column at the same time.

The gas chromatograph, and possibly the recorder also, will have a sensitivity or attenuation control (higher attenuation = lower sensitivity) which can be adjusted to vary the heights of the recorded peaks which signal the elution of the compounds. For the sake of clarity, the setting should be such that the major component of interest registers nearly full scale. The zeroing control or controls should be used to set the base line to near the bottom of the chart paper before injecting the sample.

Since the area under a VPC peak is proportional to the amount of compound eluted, the per cent composition of a mixture can be approximated by comparing relative peak sizes. This analysis assumes equal sensitivity of the detector to the different compounds, but for similar compounds, particularly with a thermal conductivity detector, it is reasonably accurate. Lacking an electronic or mechanical integrator, the simplest method for

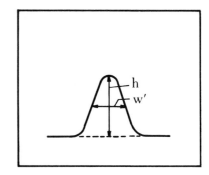

FIGURE 5.5   Measurement of peak area.

measuring the area of a peak is by triangulation (Fig. 5.5). The height (h) of the peak above the interpolated base line is multiplied by the width (w′) of the peak at half its height. Alternatively, if the peak is very asymmetric or very broad, a tracing of the peak on heavy paper can be cut out and weighed on a sensitive balance.

EXPERIMENTS

Analysis of Distillation Fractions

In order to measure the separation efficiency of your fractional distillation (p. 42), analyze the first and last fractions by VPC. A sample of approximately one microliter will be sufficient. Almost any type of column will separate benzene and toluene simply on the basis of their widely (for VPC) different boiling points. Because of the ease with which they can be separated, the temperature and flow rate are likewise not critical and should be adjusted in accordance with the general directions given. Estimate the composition of the distillation fractions by triangulation of the VPC peaks.

Effect of Operating Conditions on VPC Separation Efficiency

Using the aforementioned mixture, or better, benzene (bp 80.1°) and cyclohexane (bp 80.7°), determine the effect of sample size, flow rate, and column temperature on the resolution of a mixture by VPC. This is best done as a class project, with each person testing two or three combinations of variables and analyzing his results. The pooled results of the class can then be compared and conclusions drawn. If two or more chromatographs (or a chromatograph with two columns) are available, the effect of the liquid phase can also be observed. For judging separation efficiency, relative R_F values are used, along with the more qualitative observations of peak overlap and retention times. The latter should not be unnecessarily long, since if you were devising a routine quality check for a chemical product, the more samples you could run in a given amount of time, the better.

QUESTIONS

*1. On a column containing a polar liquid phase, in what order would hexane and ethoxyamine be eluted? Explain.

*2. Why is a flame ionization detector not usually used if the separated compounds are to be collected? Suggest how an instrument with such a detector might be modified to allow fraction collection.

*3. Using the formula on page 47 and the peak in Figure 5.4, calculate the number of theoretical plates in the column shown and also the R_F of the compound.

4. Using a chromatogram from your experiment, calculate the number of theoretical plates in the column you used. An alternative formula is $n = 5.8(d/w')^2$, where w' is the width of the peak at half its height (Fig. 5.5). This latter formula is theoretically less accurate than the one given on page 47, but it is often easier to obtain an accurate value of w' than of w. Explain why this might be so.

References

A. Keulemans, *Gas Chromatography*, 2nd Ed. Van Nostrand Reinhold, New York, 1959.
J. Schupp, *In* A. Weissberger (Ed.), *Techniques in Organic Chemistry*, Vol. XIII. Wiley-Interscience, 1968.

CHAPTER 6

THIN LAYER CHROMATOGRAPHY

It was mentioned in Chapter 5 that the term chromatography arose in another context which antedates VPC. This is solid-liquid chromatography, in which the substances to be separated are partitioned between a moving liquid phase and a solid such as silica gel ($SiO_2 \cdot xH_2O$), paper (cellulose), or alumina (Al_2O_3). These materials have large surface areas, on which exchange of molecules between the solid and a solvent flowing over the surface occurs by successive adsorption and desorption. In some cases the exchange between phases may involve partition between solvent and an immiscible adsorbed liquid (e.g., water on cellulose), but the method of operation is the same. The chromatography is carried out by applying the sample at one end of a thin layer or column (Chapt. 7) of the solid adsorbent and passing solvent over the sample and adsorbent. The components of the mixture are then detected on or eluted from the zones where they are selectively adsorbed. These two types of chromatography are completely analogous to gas chromatography, the only difference being the nature of the phases between which the compounds are partitioned.

Thin layer chromatography (TLC) is primarily a tool for rapid qualitative analysis, and it is extremely effective and convenient for this purpose. A microscopic amount of sample can be applied at one end of a small plate covered on one side with a thin adsorbent coating. The plate is then dipped into a shallow pool of solvent which rises on the coated layer by capillary action, permitting the compounds of a mixture to move with the solvent to differing heights. The individual components can then be detected as separate spots along the plate. The TLC method can be scaled up with large plates (20×100 cm) and thicker coatings of adsorbent so that gram quantities of a mixture can be resolved and the components recovered from the plate (preparative TLC).

The chief uses of TLC are to determine the number of components in a sample, to detect a given compound or compounds in a very crude mixture, and in a preliminary trial to find conditions prior to running a column

51

chromatogram. Since tiny amounts of material are exposed on an open surface, TLC is limited to relatively nonvolatile substances; in this respect it is complementary to gas chromatography. Unlike the latter, however, TLC is not easily adapted to quantitative determinations, since neither the sensitivity nor the resolving power, in general, is as high as in VPC.

To detect all of the compounds in a mixture, the developed TLC plate is treated with some general reagent such as iodine vapor to give brown spots, or H_2SO_4 followed by charring. For detection of a given compound in a mixture, the TLC behavior of the mixture and that of an authentic sample of the compound should be compared on the same plate. A given compound will move on the plate to the same extent, relative to the solvent front (R_F value), under the same conditions (sample quantity, solvent, temperature, coating). For qualitative work, however, these conditions are often not rigidly controlled, and the mobility of a compound on TLC plates run at different times is less reproducible than is the R_T value of VPC.

If a compound is colored, or can be visualized in a distinctive way by a spray reagent or by using a coating containing a fluorescent dye, this of course greatly aids in characterizing the compound in a mixture. (An analogous approach can be used in VPC if different compounds in a mixture respond with different sensitivities to different types of detectors.)

PREPARATION OF THIN LAYER SLIDES

Standard microscope slides are cheap and convenient for routine qualitative TLC. Two methods for coating TLC slides are most often used: (a) by simply dipping into a slurry of the adsorbent in organic solvent, or (b) by spreading an aqueous suspension of the adsorbent over the slides. The dipping procedure is simple, and slides can be prepared in a few seconds. Spreading is more time consuming, and the plates must then be dried at room temperature for several hours or briefly in an oven. The spreading method produces a more uniform and reproducible coating and is the only practical way to prepare larger plates. The adhesive action of the binder is more fully utilized in the aqueous preparation, and thicker coatings can be prepared.

Dipping Procedure. The slurry is prepared in the proportions 1 g silica gel:3 ml chloroform. A sufficient amount must be made up to fill the container to a depth of about 3 inches. The level must be maintained by refilling periodically, and solvent must be added as necessary to replace evaporation losses.

Shake the slurry to mix it thoroughly. Hold two clean slides together by one end, immerse in the slurry to about 1/4-inch from the top and withdraw them with a slow steady motion. (Fig. 6.1). Touch a bottom corner to the lip of the bottle to remove the excess coating and securely recap the bottle of slurry. Separate the slides and lay them on a bench, coated side up, to

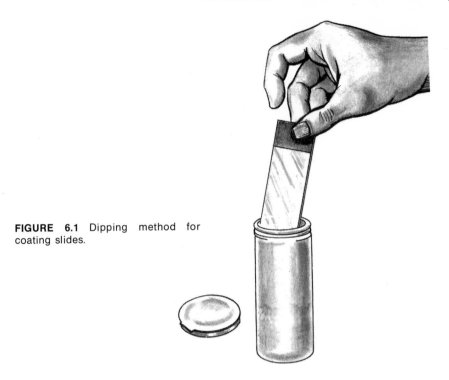

FIGURE 6.1 Dipping method for coating slides.

dry. As soon as they are uniformly white, they are ready for use. Streaked or unevenly coated plates result from incomplete mixing of the slurry. If the layers are thin and grainy, the slurry is too dilute.

Spreading Procedure. Wipe clean a section of bench top and lay out eight slides in a block 4 slides wide and 2 high, with edges touching (Fig. 6.2). Prepare a roller by placing short lengths of rubber tubing at each end of a piece of 7 mm glass rod. Weigh out 4.0 g of Silica Gel G in a 7 dram plastic cap vial. When these preparations have been made, add 8 ml of water to the silica gel and shake vigorously for 30 seconds. The aqueous slurry will thicken in a few minutes and must be spread without delay. Pour the slurry evenly across one end of the block of slides and spread it across the slides by rolling the rod across and back with a smooth motion. Excessive rolling will cause ridges at the edges of the slides.

FIGURE 6.2 Spreading method for coating slides.

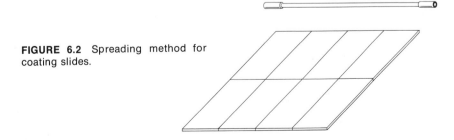

When the surface of the slides has become dull (1 to 2 minutes), push them apart with a spatula tip and slide them onto a piece of cardboard; handle carefully to avoid damaging the edge of the coating. Allow the slide to dry in air for 10 to 15 minutes. Drying can be completed in an oven at 80 to 100° for 30 minutes or at room temperature overnight.

TLC PROCEDURE

Application of Sample.   The sample is applied with a very fine capillary. These should be about one-tenth the diameter of a melting point capillary, and can be conveniently prepared by softening a 1-cm section in the middle of a melting point capillary with a microburner and drawing out to about 4 to 5 cm. Break the thin portion to make two TLC capillaries. A larger quantity of capillaries can be made from a disposable soft-glass pipet; draw out to the appropriate thickness (about a 2-foot length) and cut up the capillary section into 3-inch pieces.

The sample to be applied can be dissolved in any volatile solvent; acetone or methylene chloride is convenient. Make a roughly 5 to 10 per cent solution of the sample in a 10×75 mm test tube. Dip a capillary into the solution and then touch the end very lightly at a spot about 1 cm from the end of the TLC slide (Fig. 6.3). The spot should spread to a diameter of no more than 1 mm. If it is necessary to apply a larger amount, let the first spot dry completely and then touch the capillary again at the same place. It is important that no crystals from the sample be transferred to the plate or form when the spot is applied, since this causes streaking.

Two common errors are allowing the spot to become too broad and applying too much sample. Slides prepared by dipping usually have a thinner coating than those made by spreading, and the former therefore require smaller amounts of sample. For a preliminary investigation, it is useful to make several spots with one, two, or three applications to determine the proper volume of solution to apply. Three spots can easily be placed on a microscope slide. Four spots is the maximum, since the zones spread as the plate is developed.

FIGURE 6.3 Spotting TLC plates.

Choice of Solvents. The rate at which compounds move on an adsorbent increases with increasing polarity of the solvent in either thin layer or column chromatography. Optimum separation of closely similar compounds in TLC can usually be effected when the R_F values are 0.3 to 0.5, since the spots become diffuse as they move further. Commonly used solvents for chromatography, in order of increasing polarity, are hexane, carbon tetrachloride, benzene, ether, chloroform, ethyl acetate, acetone, and alcohols.

To determine the proper solvent, apply test spots of the sample spaced about 1 cm apart along a plate. Touch a capillary containing the solvent to be tested to the center of a spot and permit the solvent to flow out. Repeat with each solvent on a different spot. A circle about 1 cm in diameter will be formed; mark the outer edge (solvent front) around each spot and then observe the relative movement of the sample away from the center point, either directly or after staining (Fig. 6.4).

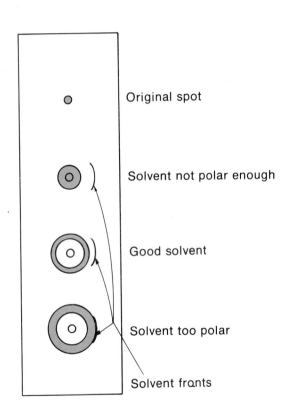

FIGURE 6.4 Testing solvents for TLC.

Chloroform is the most generally useful TLC solvent for a fairly wide range of compounds. If migration of the spots is too slow, a few per cent of methanol can be added to increase the polarity. Another approach, which is quite effective if some separation occurs with low R_F values, is to develop the plate several times, allowing it to dry between each pass.

Development and Visualization. The developing chamber is a wide-mouth bottle with snap-on plastic lid. The bottle is filled with solvent to a depth of about 0.5 cm; be sure that the level is not above the sample spots. A piece of filter paper can be fitted around the inside wall to serve as a wick for maintaining an atmosphere of solvent vapor. Lower the plate gently, spotted end down, into the chamber and replace the cap. When the solvent has risen to about 1 cm from the top of the slide, remove the plate and let it stand for a few minutes to dry completely. Mark any visible spots by pricking an outline with a capillary.

For visualization of any colorless compounds the slide is then placed in a jar containing a few crystals of iodine. Nearly all compounds except saturated hydrocarbons and halides form colored complexes with iodine vapor; these show up as brown or violet spots on the slide. The spots generally disappear in a minute or two, and they should be marked as soon as the slide is removed from the jar. The relative intensity of the spots generated by iodine is not an accurate indication of the amounts of compounds present, since the extent of complex formation varies.

EXPERIMENTS

Nitroanilines (see Chapt. 7). Prepare eight TLC plates by the method designated by your instructor, and prepare 6 to 8 TLC capillaries. Obtain in labeled 10 × 75 mm test tubes very small samples of *ortho-*, *meta-*, and *para*-nitroaniline and 2,4-dinitroaniline (about 5 mg or just enough to cover the bottom of the test tube is sufficient). Dissolve these samples in about 0.5 ml of acetone; use a little more solvent if necessary to completely dissolve the samples.

Spot on one plate samples of each of the four compounds and develop the plate with chloroform. If the spots are not distinct or are too large or have run together, repeat. Sketch in your notebook the appearance of the developed plate, and then develop again. If you feel it may be useful, develop a third time. The spots should be visible on the dry plate, but standing in iodine may bring them out more distinctly. (It is not practical to redevelop a plate that has been exposed to iodine, even if the color disappears, since some chemical change may have occurred.)

On other plates spot mixtures of two or three of the compounds in various combinations in the same lane; apply the second solution right over the spot of the first, taking care to avoid overloading. Spot the individual compounds in other lanes on the same plates. Develop and record the appearance of each plate.

When satisfactory results have been obtained with single

compounds and known mixtures, obtain samples of Nitroaniline Unknowns 1 and 2 and determine which compound(s) are present in each.

Unknowns. Obtain samples of compounds A and B and Unknown 3. One of these compounds is p-dimethoxybenzene and the other is p-methoxyphenol. Since they are both colorless, they must be visualized with iodine. Find a solvent that will permit separation of A and B, and determine whether Unknown 3 contains either or both of the compounds.

QUESTIONS

⋆1. What will be the result of the following errors in TLC technique:
   a. Too much sample applied
   b. Solvent of too high polarity
   c. Solvent pool in developing jar too deep

⋆2. Explain why the presence of crystalline material at the starting point causes streaking of the plate.

3. Comment on any differences in the relative affinities of the nitroanilines toward iodine. Can you suggest any correlation between this property and other properties of the nitroanilines (see question 5, Chapt. 7).

4. On the basis of the relative polarities expected for p-dimethoxybenzene and p-methoxyphenol, and the R_F values observed for compounds A and B, suggest the identities of A and B.

References

J. M. Bobbitt, *Thin Layer Chromatography.* Van Nostrand Reinhold, New York, 1963.

K. Randerath, *Thin-Layer Chromatography.* Academic Press, New York, 1966.

E. Stahl, (Ed.), *Thin Layer Chromatography, A Laboratory Handbook*, 2nd Ed. Springer Verlag, New York, 1969.

CHAPTER 7

COLUMN CHROMATOGRAPHY: SEPARATION OF NITROANILINES

In column chromatography, partition between a solid support and solvent provides a means of separation *and isolation* of the components in a mixture. The solid adsorbent is contained in a vertical tube, and after the sample is applied by adsorption from a small volume of solution, solvent is allowed to flow through the adsorbent column. The process develops the chromatogram into bands containing the individual compounds. These bands are then eluted (washed off) in sequence by further solvent.

This method of separation is capable of high resolution, but it can be fairly laborious. If one of the components of the mixture can be partially recovered by crystallization, it is advantageous to do this before proceeding to chromatography. The feasibility of a column separation is usually examined by TLC beforehand; the choice of the correct adsorbent and solvent system can often be made on the basis of preliminary TLC data.

In general, the amount of adsorbent is about 30 to 50 times the sample weight. For rough separation of a major component from tarry impurities, a much smaller column may be used. The column is developed using the least polar solvent that permits movement of material on the column at a practical rate. The order of polarity is essentially the same as that for TLC. It may be necessary to use a more polar solvent to effect elution of zones. When this is done, the second solvent is blended into the eluting liquid very gradually; the effect of just a few per cent of a more polar solvent may be quite marked, and care must be taken to avoid coalescing the zones that have been developed.

The development and elution steps in column chromatography are usually not entirely distinct. With colored compounds it may be possible to stop the process after separate visible bands have been developed, extrude the adsorbent column, and cut out the zones for extraction of the compounds. Normally, however, separation is not completely sharp, and the fastest moving band may begin to emerge from the column before the separation of slower moving bands is complete. With colorless mixtures, the progress of the column cannot be assessed until the eluate fractions are examined. In this case, fairly small fractions are collected and samples from fractions at suitable intervals are analyzed by TLC or other means to determine which ones are homogeneous and can be combined. As in any partition process, the first material eluted is usually quite pure, but separation of a pure component from intermediate zones may be very difficult.

PREPARATION OF COLUMN

A typical chromatography column is illustrated in Figure 7.1. The adsorbent is supported in the tube on a fritted glass disc or a plug of glass wool or cotton with a layer of sand to retain particles. The size of the tube should be chosen so that the adsorbent column is about one-half to two-

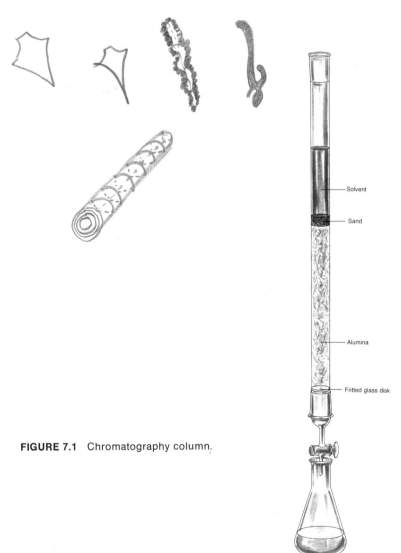

FIGURE 7.1 Chromatography column.

thirds the height of the tube, providing space for solvent above the adsorbent. The ratio of diameter to height of the column is usually 1:8–12. Fill the tube to about two-thirds the height with the initial solvent to be used in the chromatogram, and put the supporting plug and sand in place. Add the solid adsorbent in a fine stream so that it falls directly into the solvent. Alternatively, the adsorbent can be mixed with solvent and this slurry transferred to the tube and allowed to settle. Drain out the solvent at a rate to maintain the original level as the adsorbent is added. As the adsorbent settles, tap the tube gently and continuously to produce a compact, uniform column of adsorbent. While filling and using the column, *always maintain the solvent level above the top of the adsorbent column.* If the top of the column becomes dry, channels will form and greatly reduce the effectiveness of the chromatogram.

After the adsorbent has settled, pour in a small layer of sand to protect the top layer of adsorbent from being disturbed, and drain the solvent to within about 0.5 cm of the top of the sand.

RUNNING THE CHROMATOGRAM

The sample to be chromatographed is dissolved in the *minimum* volume of solvent so that the sample is adsorbed in a very narrow zone. Preferably the solvent is the one used to prepare the column. If the solubility is less than 10 to 20 per cent (1 to 2 ml for 0.2 g), a more powerful (usually more polar) solvent must be used. Allow solvent to flow slowly out of the stopcock until the sample has drained completely onto the adsorbent.

The space above the adsorbent column is now filled with solvent and chromatography is begun. If distinct visible bands separate as the column is developed, these should be collected in separate fractions. The solvent polarity is increased if necessary to keep material moving on the column. If no bands appear, fractions of a convenient arbitrary size are collected, usually about 50 ml per gram of sample. The samples are then examined by TLC or by evaporation, and similar fractions are combined.

PREPARATION AND SEPARATION OF *ORTHO-* AND *PARA*-NITROANILINE

The preparation of nitroaniline illustrates a typical electrophilic aromatic substitution and the use of a removable protecting group in synthesis. Aniline is itself so reactive that direct nitration leads to tars. Acetylation of aniline, as carried out in Chapter 2, reduces the reactivity and permits mononitration of the ring under carefully controlled conditions; subsequent acid hydrolysis regenerates the amines.

	ortho- nitroaniline	para- nitroaniline	2,4-dinitro- aniline
Melting point	72°	147°	188°
Solubility in ethanol	30%	7%	0.7%

This process provides a practical source of p-nitroaniline, which is formed in 75 to 80 per cent yield. The amounts of *ortho* isomer and dinitro-aniline are 10 to 15 per cent and 5 to 10 per cent, respectively; the *meta* isomer is negligible. This is quite a favorable situation for isolation of products, since it is seen that the compound present in largest amount is higher melting and considerably less soluble than the next most abundant product. This means that most of the *para* isomer can be obtained in quite good purity by crystallization of the mixture.

Separation of the remaining mixture and isolation of the *ortho* isomer by further crystallization would be very inefficient, however, since the less soluble minor components (*para* isomer and dinitro compound) would probably cocrystallize with the predominant *ortho* isomer. On the other hand, chromatography is quite suitable for a mixture of this type, since the less soluble, higher melting minor components will be retained more firmly on the absorbent. Although the two nitroanilines have the same functional groups, they differ significantly in polarity, as reflected in the large difference in physical properties.

In this experiment, the preparation of p-nitroaniline will be carried out, and the *ortho* isomer will be isolated also as a demonstration of column chromatography. Alternatively, your instructor may provide a sample of a typical mixture that would be obtained in this reaction after the bulk of the *para* isomer was removed.

EXPERIMENTAL PROCEDURE

In a 125 ml Erlenmeyer flask place 2.7 g of acetanilide from your preparation (Chapt. 2) or from your instructor. If your own sample is used, be sure it is completely dry before weighing. Obtain 9 ml of concentrated sulfuric acid (10 ml graduated cylinder), pour about half of the acid into the flask and dissolve most of the acetanilide by swirling and stirring (a small amount of solid remaining will subsequently dissolve). Place the flask in an ice bath (large beaker). Measure out accurately 1.5 ml of concentrated nitric acid and add it to the remaining sulfuric acid. Mix the acids by drawing up samples from the bottom with a transfer pipet and bulb and emptying them on top; complete the mixing by discharging a stream of bubbles from the empty pipet at the bottom of the cylinder.

Using the pipet and bulb, add the mixed acid in small portions (about 0.5 ml) to the cooled sulfuric acid solution of acetanilide. Swirl the flask in the ice bath after each portion is added. The flask should not become perceptibly warm to the touch; the addition requires 10 to 15 minutes. After 20 minutes, including addition time, add 25 ml of a mixture of ice and water (loosely pack a 16×150 mm test tube with ice and fill with water) to the reaction mixture.

The resulting suspension of nitroacetanilides is now hydrolyzed in the same flask, using the aqueous sulfuric acid. Add a boiling stone and heat the flask on a wire gauze with a burner flame. Watch carefully as the color darkens and the solid begins to dissolve, and remove the flame and decrease it if necessary to avoid excessively vigorous boiling and foaming. Continue heating at a gentle boil for 15 minutes and then cool the flask and contents in an ice bath. Place the bath and flask in the hood and add, in 5 or 6 portions, 25 ml of concentrated aqueous ammonium hydroxide; swirl in the ice bath after each addition.

Collect the precipitated nitroaniline in a Büchner funnel by suction filtration. Rinse the flask with 3 ml of water, disconnect the suction tubing, pour in the rinse, and then suck the wash water through the precipitate, press the solid, and allow air to be drawn through for 2 or 3 minutes.

Transfer the yellow crystals, which can be peeled cleanly from the filter paper, to a 25×100 mm test tube and add 4 ml of ethanol. Heat and stir the mixture on the steam bath until the alcohol boils and most of the solid dissolves, then cool in an ice bath for 10 to 15 minutes to crystallize the p-nitroaniline. Prepare a suction filtration set-up such as that in Figure 2.3, to collect the filtrate in a tared receiver. An 18×150 mm test tube supported inside a 500 ml suction flask is quite satisfactory for this purpose. Collect the p-nitroaniline crystals by suction filtration in a Hirsch funnel, draining the dark viscous mother liquor into the tared test tube. Transfer the crystals as completely as possible; use only a minimum volume (about 0.5 ml) of ethanol to rinse and wash. Spread the p-nitroaniline on glassine paper to dry, and run a TLC on both the crystals and the oily filtrate [see Chapt. 6]. Record the TLC results and, after the crystals are dry, the weight and melting point of the p-nitroaniline.

Para-Nitroaniline.   For final purification, the p-nitroaniline obtained usually requires recrystallization. If time permits, dissolve the crude product using 15 ml of ethanol per gram, warming if needed to dissolve. Add charcoal, swirl for 1 to 2 minutes, and filter through a fluted paper. Refilter if necessary to obtain a clear yellow solution. The relatively large volume of ethanol is necessary to avoid crystallization of the product in the filter during removal of the charcoal; the filtrate should be concentrated to about one-third the initial volume to minimize loss in the final crystallization. Record melting point, TLC behavior, weight, and percentage yield of the recrystallized product and submit the sample to your instructor.

o-Nitroaniline.   Using the steam bath and aspirator vacuum (Fig. 3.7), evaporate the alcohol and water from the mother liquor from the initial crystallization of the *p*-nitroaniline and record the weight of the crude residue.

Prepare a chromatography column with 15 g of alumina, using chloroform as solvent. A 10 × 300 mm chromatography tube or a 25 ml pinch-cock buret can be used for the column.

Dissolve the nitroaniline residue in a minimum volume of chloroform; 1 ml should be plenty. (If a crude *o*-nitroaniline preparation is provided instead, weigh out 0.25 g and proceed in the same way.) Apply the sample to the absorbent, then develop and elute the column with chloroform. Collect the eluate in test tubes, taking 5 ml fractions beginning with the first colored solution that emerges from the column. After five 5-ml fractions, collect one final large fraction until either all of the colored material is eluted or the volume of this fraction reaches 25 ml. Examine each fraction by TLC (three fractions per plate) and pool fractions on the basis of TLC behavior; if two fractions show the same TLC spot(s), combine them.

Evaporate and weigh any fractions or groups of fractions that contain a single component. Compare the TLC with appropriate authentic samples; determine the melting point and weight of the crystalline residues. Record the yield of *ortho*-nitroaniline and any other components isolated from the chromatogram.

QUESTIONS

⋆1. Why is it important to apply the sample to the column in the smallest volume of solution possible? What would be the effect of using an excessive amount of solvent in this step?

⋆2. Chromatography can be carried out by packing the column with Celite (mineral earth) that is saturated with water which contains a few per cent of *n*-butanol. After applying the sample, a mixture of chloroform and butanol is passed through the column to effect separation. What type of process is actually being carried out on the column? What is the function of the butanol? Why must the butanol be present in both water and chloroform?

★3. How and why is the efficiency of the column reduced by a channel or dry spot?

4. Suggest a reason for the fact that *meta*- and *para*-nitroaniline are very similar in physical properties and both differ markedly from the *ortho* isomer. Which of the three compounds would you expect to be the most volatile?

5. How does the chromatographic behavior of the nitroanilines on alumina and silica gel compare?

References

J. M. Bobbitt, *Introduction to Chromatography*. Von Nostrand Reinhold, New York, 1968.
J. G. Kirchner, *In* E. S. Perry and A. Weissberger (Eds.), *Techniques of Organic Chemistry*. Interscience, New York, 1967.

STEAM DISTILLATION OF ESSENTIAL OILS

In Chapter 4 it was seen that distillation of a mixture (more accurately, solution) of two miscible liquids depended on the vapor pressure and mole fraction of the two components. A different situation is encountered in the distillation of a mixture of two compounds which are not mutually soluble. In this case, the total vapor pressure above the mixture is given simply by Dalton's law, i.e., it is the sum of the vapor pressures, $P_T = P_A^\circ + P_B^\circ$, *independent* of the amounts of the two compounds. Thus as long as some of each liquid phase is present, the distillate will have a constant composition, and the boiling point will be lower than that of either pure component.

Steam distillation, in which one of the immiscible components is water, provides a means of distillation of a slightly volatile compound at a temperature far below its atmospheric boiling point. It is also useful for the separation of any organic compound that is contaminated with a large amount of nonvolatile ballast. In this case, simple distillation even in vacuum is ruled out because destructively high temperatures are required to distill the relatively small amount of volatile material; mechanical entrapment will prevent its complete removal. Examples of this application are the recovery of oils from plant gums or resins. Extraction of the plant material with solvent removes fats and other nonvolatile constituents as well as the volatile oils; the latter can then be selectively separated by steam distillation.

65

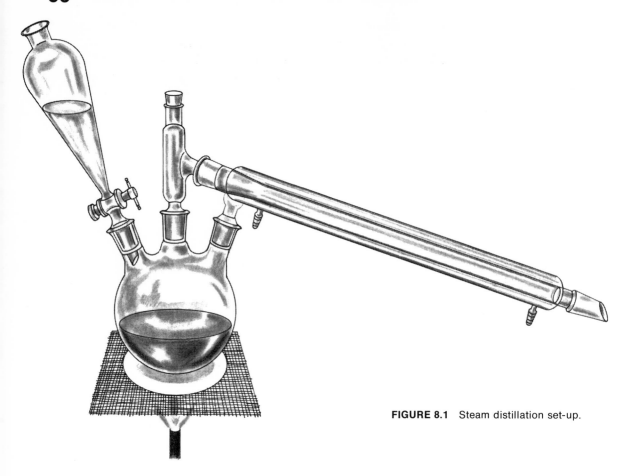

FIGURE 8.1 Steam distillation set-up.

APPARATUS AND TECHNIQUE

Steam distillation is most conveniently carried out in a 2- or 3-neck round-bottom flask. A typical set-up is illustrated in Figure 8.1. The top of the distilling head and the extra neck are sealed off with a cork or stopper; the boiling point is usually of no interest. Steam is provided by addition of water and sufficient heat from a burner or heating mantle. Alternatively, steam can be supplied from an external boiling flask or a steam line on the desk; this is necessary to provide agitation when steam distilling a tarry or very bulky mass. When using an external source of steam, a trap must be included between the source of steam and the distilling flask in case a drop in steam pressure causes the contents of the flask to be sucked back through the steam inlet.

ESSENTIAL OILS

The characteristic aromas of plants are due to the volatile or essential oils, which have been used since antiquity as a source of fragrances and flavorings. These oils occur in all living parts of the plant; they are often concentrated in twigs, flowers, and seeds. Essential oils are generally complex mixtures of hydrocarbons, alcohols, and carbonyl compounds, mostly belonging to the broad group of plant products known as terpenes. Some terpenes such as borneol (see Chapt. 19) and a few hydrocarbons are found in many different plant species. Certain other compounds, particularly aromatic aldehydes and phenols, are a major constituent in the essential oils of one or a few plants and impart the characteristic aroma of condiments such as cloves, cinnamon, or vanilla.

These essential oils are usually isolated from the plant tissue by steam distillation; individual compounds are then separated by crystallization, fractional distillation, or chromatography. In this experiment, the essential oil will be steam distilled from one of the three widely used spices—anise, cumin, and cloves, and the major constituent will be characterized as a derivative.

EXPERIMENTAL PROCEDURE

Place 10 g of ground anise or cumin seeds, or clove buds in a 500 ml 3-neck flask and add 150 ml of water. Assemble the set-up as in Figure 8.1 with a 125 ml Erlenmeyer flask as the receiver. Fill the dropping funnel with water. Heat the mixture until steady distillation begins and then add water from the funnel at a rate to maintain the original level. Collect about 100 ml of distillate.

Extract the distillate in a separatory funnel with two 10-ml portions of methylene chloride. If the methylene chloride layers are separated carefully, drying is unnecessary. Evaporate the methylene chloride on the steam bath, transfer to a tared test tube, concentrate to an oily residue, and weigh and calculate the yield of oil based on the weight of plant material. Continue with the appropriate following procedures.

Anise Oil. The essential oil of anise (*Pimpinella anisium*) is predominantly a single compound, *p-trans*-propenylanisole (anethole). This ether has a melting point slightly below room temperature, and the compound can be crystallized by chilling the oil in an ice bath. Since the anethole is only 80 to 90 per cent pure at this stage and is too low melting for convenient manipulation, it is converted to the dibromide for characterization.

Anethole, mp 22°   Anethole dibromide, mp 65°

To prepare the dibromide, dissolve the crude anethole in 1 ml of pentane and chill in an ice bath (small beaker). To this solution, add dropwise, using a transfer pipet and bulb, a 10 per cent solution of bromine in pentane until a red color persists; about 2 ml of bromine solution is required. Concentrate the solution on the steam bath to about 1 ml volume. A few oily red droplets may separate at this point. Transfer the pentane solution with a pipet to a clean 10 × 75 mm test tube, chill, and scratch to crystallize. Collect the dibromide and recrystallize, using about 1/2 to 1 ml of pentane. Determine the melting point and compare with that of anethole dibromide.

Cumin Oil. The major volatile constituent of the seeds of cumin (*Cuminum cyminum* L.) is *p*-isopropylbenzaldehyde (cuminaldehyde). However, this compound contributes only part of the aroma of the essential oil, as will be observed by comparing the odor of the oil before and after isolating this aldehyde as a semicarbazone derivative.

Cuminaldehyde
semicarbazone, mp 216°

To prepare the semicarbazone, dissolve 0.20 g of semicarbazide hydrochloride and 0.30 g of anhydrous sodium acetate in 2 ml of water and add 3 ml of ethanol. Add this solution to the cumin oil, warm briefly on the steam bath, cool, and allow the derivative to crystallize. Collect the crystals in a Hirsch funnel and recrystallize from methanol. Compare the melting point to that reported for cuminaldehyde semicarbazone; recrystallize again if necessary.

Clove Oil. Oil of cloves (*Eugenia caryophyllata*) is rich in 4-allyl-3-methoxyphenol (eugenol); other compounds, including the sesquiterpene caryophyllene, are present in very small amounts. For characterization, the eugenol can be converted to the benzoate ester.

Eugenol benzoate, mp 70°

Place about 0.2 ml of the oil in a test tube and add 1 ml of water. Add 1*N* KOH or NaOH drop by drop until the oily layer dissolves; a completely clear solution will not result, but there should be no oily droplets. To this solution, add about 0.1 ml (4 to 5 drops) of benzoyl chloride; avoid an excess of this reagent. Stir and warm the mixture on the steam bath for about 5 minutes, then cool until the oil becomes gummy and then solidifies. Rub the solid until it is granular; if a gummy consistency persists, decant most of the aqueous layer and add a few drops of methanol. Collect the solid on a Hirsch funnel and wash with water on the filter. Recrystallize the moist solid from a minimum volume of methanol; if an oil begins to separate, warm slightly and rub with a seed while the solution cools again. Collect the recrystallized solid, dry in air, and record the melting point.

QUESTION

★The composition of the distillate in the steam distillation of a substance X depends directly on the partial pressures of water and X, and these are in turn proportional to the number of moles of each component in the vapor phase. It is possible to make a rough determination of the molecular weight of X from the vapor pressure and the weight of water and the weight of X in a portion of the steam distillate; thus,

$$\frac{moles_{H_2O}}{moles_X} = \frac{P^\circ_{H_2O}}{P^\circ_X}$$

In the distillation of X, 1.0 g of X and 4.0 g of water distilled at a temperature of 99° ($P^{\circ}_{H_2O}$ at 99° = 733 mm). Calculate the molecular weight of X. What is the major source of error in this determination?

Reference

E. Guenther, *The Essential Oils*. Van Nostrand, New York, 1949–1952. (In six volumes.)

CHAPTER 9

INFRARED ABSORPTION SPECTROSCOPY

INTRODUCTION

As implied by the name, infrared absorption spectroscopy is the measurement of the amount of radiation absorbed by compounds within the infrared region of the electromagnetic spectrum. Specifically, this is radiation with wave-lengths between 2.5 and 15μ ($1\mu = 1$ micron $= 10^{-4}$ cm) and frequencies between 4000 and 650 cm^{-1} (cm^{-1} = cycles per cm = wavenumbers = "reciprocal centimeters"). As with other types of absorption, the molecules in the compound are excited to a higher energy state as a result of the radiation absorbed. Infrared radiation absorption causes energy changes on the order of 2 to 10 kcal/mole, and the excited state is one involving a greater amplitude of molecular vibration. Each absorption band or peak in the spectrum corresponds to the excitation of a different type (mode) of vibration of the atoms, and the positions of these peaks (measured in μ or cm^{-1}) provide useful information about the structure of the molecule.

For simple molecules containing few atoms, the number and position of peaks in the infrared spectrum can be calculated, and the spectrum can be completely analyzed. The major use of infrared spectra in organic chemistry, however, depends on empirical correlations of band positions with structural units. These correlations have been derived from spectra of a large number of compounds of known structure. By using this information, the infrared spectrum of a new compound provides data on the presence or absence of certain structural features in the molecule.

Two types of vibrations, stretching and bending, are responsible for most of the peaks of importance in identification of organic compounds. A few of the types of vibrations observed are shown in Figure 9.1 using $>$CH$_2$ as a typical group. Also shown are the approximate frequencies of the radiation absorbed in exciting each type of vibration. When expressed

71

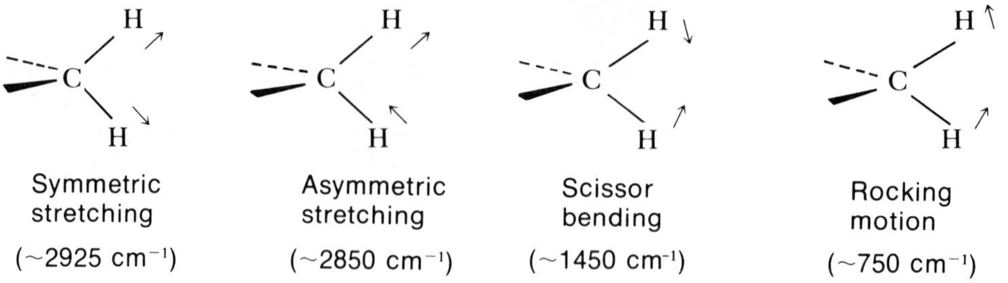

Symmetric stretching	Asymmetric stretching	Scissor bending	Rocking motion
(~2925 cm⁻¹)	(~2850 cm⁻¹)	(~1450 cm⁻¹)	(~750 cm⁻¹)

FIGURE 9.1 Normal modes of CH_2 vibration.

in cycles per second, these are also the vibrational frequencies of the indicated motions. The vibrational frequency, and thus the frequency of radiation absorbed, is determined by the force constant for the deformation (i.e., the rigidity or strength of the bond) and the masses of the atoms which are involved in the vibration: Specifically, the stronger the bond and the lighter the atoms, the higher the vibrational frequency. Bending motions are easier than stretching motions, so that the former absorb at lower frequencies. These trends are illustrated in Table 9.1 which lists the types of vibrations which appear in various regions of the infrared spectrum.

TABLE 9.1

FREQUENCY (cm⁻¹)	VIBRATION
3600–2700	X—H single bond stretching: O—H, N—H, C—H
3300–2500	Hydrogen-bonded O—H----X stretching
	Ammonium ion $\overset{+}{>}$N—H stretching
2400–2000	Triple bond and cumulated double bond stretching: C≡C, C≡N, N═C═O, C═C═O, N═C═N
1850–1550	Double bond stretching: C═O, C═N, C═C
1600–650	Single bond bending: NH_2, CH_3, C—C—C
	Single bond stretching: C—C, C—O, C—N,

Peaks in the first four regions listed are largely due to vibrations of specific types of bonds, and these are by far the most useful in compound classification. A few of the peaks appearing below 1500 cm⁻¹ are characteristic of certain functional groups, but most of the absorption bands in this region are associated with vibrations of larger groups or the molecule as a whole. Exceptions to this generalization are discussed below. This last region is often called the "fingerprint region" because of the large number of bands usually present, and since its appearance is unique for any compound, it is useful in comparing two samples for possible identity.

STRUCTURAL GROUP ANALYSIS

As mentioned, extensive correlations exist between absorption peak positions and structural units of organic molecules. In Figure 9.2 the most useful of these are summarized in the form of a chart which is scaled as a typical spectrum might be. In this chart and all infrared spectra in this book, the horizontal scale is linear in wavenumbers with the exception of a change of scale at 2000 cm^{-1}; the region of 2000 to 650 cm^{-1} is expanded two-fold relative to the range 4000 to 2000 cm^{-1}. In spectra to be presented later, the vertical scale is per cent transmittance, 100 per cent being the top of the spectrum, 0 per cent the bottom. Thus absorption of radiation at a certain frequency results in a decrease in transmittance and appears as a dip in the curve (Figs. 9.3 to 9.7).

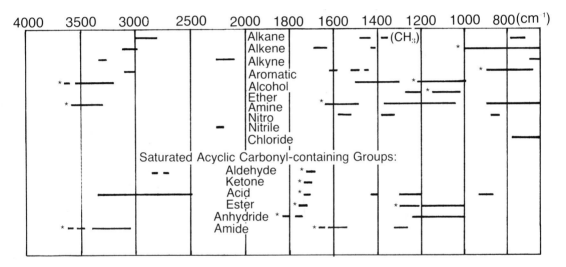

FIGURE 9.2 Infrared spectral correlation chart. *See text for discussion of this region.

For the more common structural units shown in Figure 9.2 more specific assignments and deductions may often be made. Some of these are discussed in the following paragraphs. For even more specific correlations and for similar data on other functional groups any of several books on this subject can be consulted.

Alkanes. The most prominent peaks in infrared spectra of saturated hydrocarbons and from saturated portions of more complicated organic compounds are those due to C—H stretching and bending (Fig. 9.1). The symmetric and asymmetric stretching frequencies are in the regions 3000 to 2900 and 2900 to 2800 cm^{-1}, respectively, and are usually rather weak. Since most organic compounds contain several —CH$_3$, $>$CH$_2$, and/or

$>$CH groupings, it is seldom possible to resolve these sharp absorption peaks completely, and the presence or absence of peaks in this region is simply taken to indicate the presence or absence of aliphatic C—H bonds in the molecule. The major bending modes of CH_2 and CH_3 groups appear in the 1470 to 1420 cm^{-1} and 1340 to 1380 cm^{-1} regions. The exact positions of these as well as the stretching frequencies depend upon the nature of adjacent atoms, but interpretation is usually complicated by the presence of several bands of this type or, in the case of the bending region, additional bands from other sources. The usually large number of different C—C bonds in an organic molecule makes the C—C stretching vibrations in the 1300 to 1100 cm^{-1} region uninterpretable in most cases (Fig. 9.3.).

Alkenes. Olefinic C—H stretching peaks generally appear in the region 3100 to 3000 cm^{-1}, thus differentiating saturated and unsaturated hydrocarbons. Of the C—H bending modes, the out-of-plane vibrations in the 1000 to 650 cm^{-1} region are often useful in predicting the substitution pattern of the double bond. This is illustrated in Table 9.2. The C=C stretching frequency in the 1600 to 1675 cm^{-1} region also varies with substitution but to a lesser degree. Figure 9.3 illustrates the out-of-plane bending and C=C stretching absorptions of a simple olefin in which these bands are exceptionally prominent.

TABLE 9.2   OLEFINIC C—H OUT-OF-PLANE BENDING FREQUENCIES

OLEFIN	FREQUENCY RANGE (cm^{-1})
R—CH=CH$_2$	1000–960 and 940–900
R$_2$C=CH$_2$	915–870
trans-RCH=CHR	990–940
cis-RCH=CHR	790–650
R$_2$C=CHR	850–790

Alkynes. The C—H stretching vibration of terminal acetyenes generally appears at 3300 cm^{-1} as a strong sharp band. The C≡C stretching band is found in the region 2150 to 2100 cm^{-1} if the alkyne is monosubstituted (Fig. 9.8) and at 2270 to 2150 cm^{-1} if disubstituted. The latter are usually quite weak absorptions.

Aromatic Rings. Aromatic C—H stretching absorption appears in the region 3100 to 3000 cm^{-1}. This fact taken with the corresponding frequencies of aliphatic and olefinic C—H stretching (*vide supra*) allows a reliable determination of the types of carbon-bound hydrogen in the molecule. Aromatic C—H out-of-plane bending bands in the 900 to 690 cm^{-1} region are reasonably well determined by the substitution pattern of the benzene ring as indicated in Table 9.3. In the absence of other interfering absorptions, such as those from nitro groups, these strong, usually sharp bands can be used in distinguishing positional isomers of substituted benzenes. Sharp peaks at ~ 1600 and ~ 1500 cm^{-1} are very characteristic

TABLE 9.3 *AROMATIC (PHENYL) C—H OUT-OF-PLANE BENDING FREQUENCIES*

SUBSTITUTION	FREQUENCY RANGE (cm^{-1})
mono	775–730 and 710–690
1,2-di	765–730
1,3-di	800–750 and 710–690
1,4-di	840–800
1,2,3-tri	800–760 and 740–700
1,2,4-tri	880–860 and 820–800

of all benzenoid compounds; a band at 1580 cm^{-1} appears when the ring is conjugated with a substituent (Figs. 9.5 and 9.7).

Alcohols, Phenols, and Enols. The very characteristic infrared band due to O—H stretching appears at 3650 to 3600 cm^{-1} in dilute solutions. In spectra of neat liquids or solids intermolecular hydrogen bonding broadens the band and shifts its position to lower frequency (3500 to 3200 cm^{-1}) (Fig. 9.4). Intramolecular hydrogen bonding (to $\diagdown C{=}O$, —NO_2 groups) as in enols lowers the frequency and broadens the absorption even more. Strong bands due to O—H bending and C—O stretching are observed at 1500 to 1300 cm^{-1} and 1220 to 1000 cm^{-1}, respectively. In simple alcohols and phenols the exact position of the latter is useful in classification of the hydroxyl group. Within the range given, typical frequencies observed are phenols > tertiary > secondary > primary. Further branching and unsaturation also affect the frequency somewhat.

Ethers. The asymmetric C—O stretching absorption of ethers appears in the region 1280 to 1050 cm^{-1}. As in alcohols, the exact position of this strong peak is dependent on the nature of the attached groups. Phenol and enol ethers generally absorb at 1275 to 1200 cm^{-1}, dialkyl ethers at 1150 to 1050 cm^{-1}. Epoxides have three characteristic absorptions in the 1270 to 1240, 950 to 810, and 850 to 750 cm^{-1} regions of the spectrum.

Amines. Primary and secondary amines show N—H stretching vibrations in the 3500 to 3300 cm^{-1} region (Fig. 9.10). Primary amines generally have two bands approximately 70 cm^{-1} apart due to asymmetric and symmetric stretching modes. Secondary amines show only one band. Inter- or intramolecular hydrogen bonding broadens the absorptions and lowers the frequency. In general the intensities of N—H bands are less than of O—H bands. The N—H bending and C—N stretching absorptions are not as strong as the corresponding alcohol bands and occur at approximately 100 cm^{-1} higher frequencies. In addition, NH_2 groups give an additional broad band at 900 to 700 cm^{-1} due to out-of-plane bending.

Nitro Groups. Due to the high polarity of the N—O bonds in nitro groups, the absorptions of the N—O stretching frequencies are very strong. The asymmetric stretching band appears at 1600 to 1500 cm^{-1} and the symmetric stretching at 1390 to 1300 cm^{-1} (Fig. 9.5). An additional sharp band at ~870 cm^{-1} may complicate the assignment of the aromatic substitution pattern in this region.

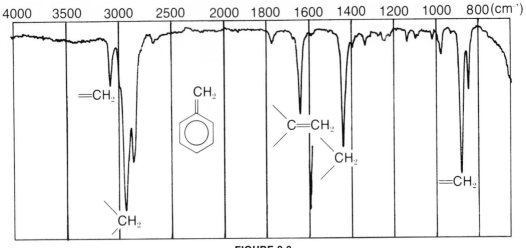

FIGURE 9.3

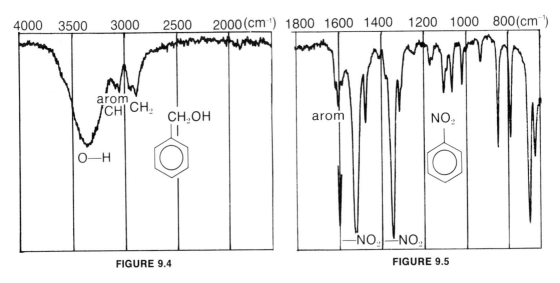

FIGURE 9.4

FIGURE 9.5

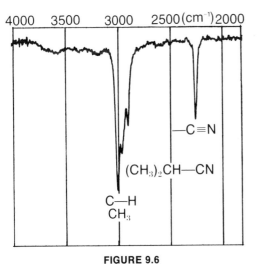

FIGURE 9.6

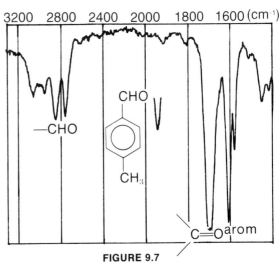

FIGURE 9.7

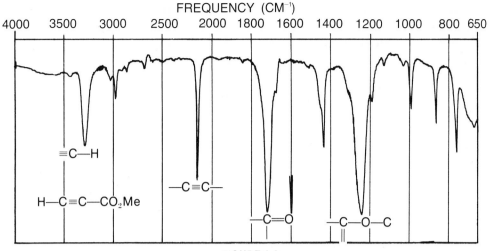

FIGURE 9.8

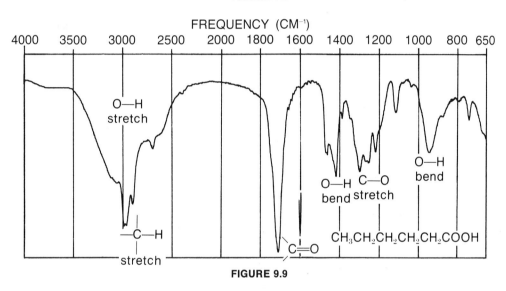

FIGURE 9.9

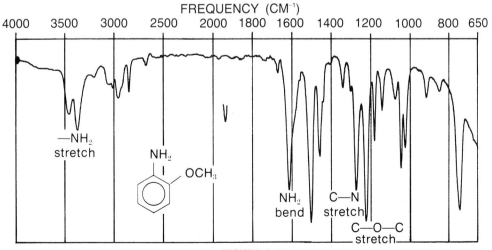

FIGURE 9.10

Nitriles. A sharp, usually strong absorption at 2260 to 2220 cm^{-1} due to C≡N stretching is very characteristic of nitriles (Fig. 9.6).

Aldehydes and Ketones. The C=O stretching frequencies of saturated aldehydes and acyclic ketones are observed at 1735 to 1710 cm^{-1} and 1720 to 1700 cm^{-1}, respectively. Adjacent unsaturation lowers the frequency by 25 to 50 cm^{-1}. Thus aryl aldehydes generally absorb at 1700 to 1690 cm^{-1} and diaryl ketones at 1670 to 1660 cm^{-1}. Intramolecular hydrogen bonding to the carbonyl oxygen also lowers the frequency by 25 to 50 cm^{-1}. Aldehydes are also recognizable by the C—H stretching vibration which appears as two peaks in the 2850 to 2700 cm^{-1} region (Fig. 9.7). The former may be obscured by aliphatic C—H stretching but the latter is usually quite prominent. Cyclic ketones (4- or 5-membered rings) absorb at higher frequencies (1780, 1745 cm^{-1}, respectively).

Carboxylic Acids. The most characteristic absorption of carboxylic acids is a broad peak extending from 3300 to 2500 cm^{-1} due largely to hydrogen bonded O—H stretching (Fig. 9.9). The C—H stretching vibrations appear as small peaks on top of this band. The carbonyl group of aliphatic acids appears at 1730 to 1700 cm^{-1} and is shifted to 1720 to 1680 cm^{-1} by adjacent unsaturation.

Carboxylic Esters and Lactones. Saturated ester carbonyl stretching is observed at 1740 to 1720 cm^{-1}. Unsaturation adjacent to the carbonyl group lowers the frequency by 10 to 15 cm^{-1} (Fig. 9.8), while unsaturation adjacent to the oxygen (enol and phenol esters) increases the frequency by 20 to 30 cm^{-1}. The C—O—C stretching of esters appears as two bands in the 1280 to 1050 cm^{-1} region. The asymmetric stretching peak at 1280 to 1150 cm^{-1} is usually strong and varies with substitution much as the corresponding ether band. Cyclic esters (lactones), like cyclic ketones, absorb at higher frequencies as the ring size decreases.

Anhydrides. Acid anhydrides are readily recognized by the presence of two high frequency (1830 to 1800 cm^{-1} and 1775 to 1740 cm^{-1}) carbonyl absorptions. As with other carbonyl stretching vibrations, the frequency is increased by incorporating the group in a ring and decreased by adjacent unsaturation. Cyclic anhydrides differ from acyclic anhydrides also in that the lower frequency band is stronger in the former, while the reverse is true of the latter (Fig. 14.1).

Amides and Lactams. Amide carbonyl stretching is observed in the 1670 to 1640 cm^{-1} region. In contrast to other carbonyl groups, both adjacent unsaturation and ring formation (lactams) cause the absorption to shift to higher frequencies. Primary and secondary amides also show N—H stretching at 3500 to 3100 cm^{-1} (see discussion of amines), and N—H bending at 1640 to 1550 cm^{-1}.

PROCEDURES FOR OBTAINING INFRARED SPECTRA

Liquid Samples. If the compound whose spectrum is to be measured is a liquid, the simplest method for mounting the sample consists of placing a thin film of the liquid between two transparent windows. The most common material used for the windows is NaCl, which is transparent throughout the normally used region of the infrared spectrum (10,000–650 cm^{-1}). Large polished single crystals are used, and it is important to remember when handling them that NaCl is soluble in water. They should be picked up *only* by the edges, preferably with gloves, to avoid marring the polished surface with moisture from your fingers.

For mounting the salt plates with the liquid sample between them, the holder illustrated in Figure 9.11 is used. The bottom metal plate is placed on a flat surface and one of the rubber gaskets is placed around the opening in the plate. This serves to cushion the relatively fragile salt plate which is placed on top of it. A drop of the liquid compound is placed in the center of the lower window, and the second salt plate is placed carefully on top, spreading the drop into a thin film. The other rubber gasket and the face plate are then added to the top of the sandwich, and the entire stack is held together by the thumb nuts on the threaded studs shown. All four nuts should be firmly tightened but not excessively so. The assembly is placed in the holder provided on the instrument.

After obtaining the spectrum, disassemble the cell and rinse the windows well with a dry, volatile solvent (CHCl$_3$, hexane, and the like) and store them in a desiccator to protect them from moisture in the air.

Sealed cells are available to contain volatile samples or solutions of compounds in volatile solvents. Solution spectra in nonpolar liquids are useful in minimizing intermolecular association of polar groups in the molecule.

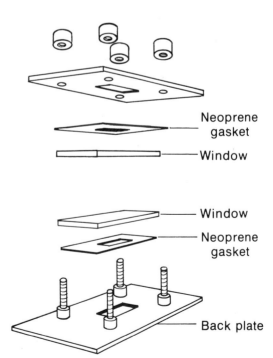

Neoprene gasket

Window

Window

Neoprene gasket

Back plate

FIGURE 9.11 Infrared salt plate assembly.

Solid Samples.   As an alternative to measuring the spectrum of a liquid solution of a solid compound, a solid solution or dispersion in KBr is usually more convenient. Approximately 1 to 2 mg of the compound is mixed with 50 to 100 mg of dry KBr powder. The mixing must be complete and vigorous enough to reduce the particle size below that which will diffract or scatter the light (less than two microns). This can be accomplished by grinding the mixture with an agate mortar and pestle or by using the pounding action of a miniature ball mill. The latter technique will be described here.

Weigh the specified quantities of material into the plastic vial provided and add a glass ball which fits loosely into the vial. Place the stoppered vial in the shaker and mix for 2 to 3 minutes at full speed. Remove the ball from the vial and loosen the powder from the sides of the vial by tapping or by scraping with a spatula.

The powder is formed into a transparent pellet by pressure in a small die. The die used here consists of a threaded metal block and two bolts with polished end surfaces. To use the die, screw one of the bolts in 5 or 6 turns, so that 1 or 2 threads remain showing. Pour the powder into the open end and distribute it evenly over the end of the bolt by tapping gently. Screw the other bolt down on top of the sample and tighten the bolts securely with the wrenches provided. Let the die stand for about a minute, during which time the KBr flows to fill the empty spaces between the original particles.

Loosen the bolts with the wrenches and remove them leaving the KBr as a window in the middle of the die. If the pellet is very cloudy, either the compound was not ground well or the bolts were not tightened enough, and a poor spectrum will result due to light scattering. Slight scattering due to a translucent pellet can be partially compensated for with an attenuating device in the reference beam of the instrument. In the absence of a commercial attenuator, a piece of wire screen works adequately to balance the lowered transmittance of the pellet.

Recording the Spectrum.   Your instructor will describe the method of operating the infrared spectrophotometer. Care should be taken in placing the chart paper as indicated so that the scale will accurately measure peak positions. It is advisable to check this positioning after the spectrum has been run by tracing one of the sharp lines in the spectrum of a polystyrene film provided. Either 1944 cm^{-1} or 1601 cm^{-1} may be used depending on the location of bands in the spectrum of the compound run (Figs. 9.3 to 9.7).

QUESTION

Figures 9.12 to 9.14 are the infrared spectra of three compounds: A ($C_{10}H_{12}O$), B ($C_{10}H_{12}O_2$), and C ($C_{10}H_{12}O$). Interpret the spectra for functional groups, and if possible suggest the

FREQUENCY (CM⁻¹)

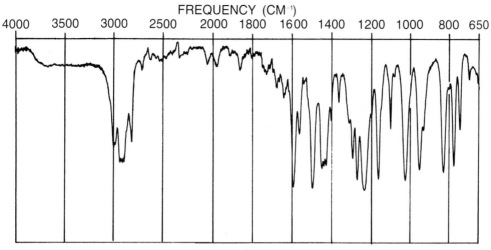

FIGURE 9.12   Compound A.

FREQUENCY (CM⁻¹)

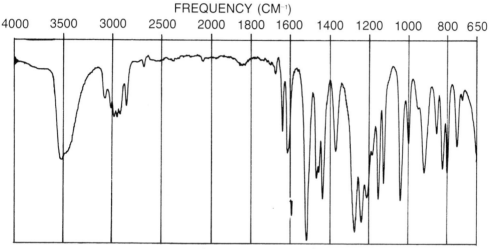

FIGURE 9.13   Compound B.

FREQUENCY (CM⁻¹)

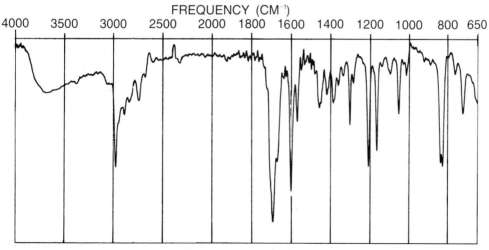

FIGURE 9.14   Compound C.

structure of each compound. The nmr spectra of these compounds are given at the end of Chapter 10.

References

Interpretation of Spectra — General

J. R. Dyer, *Applications of Absorption Spectroscopy of Organic Compounds.* Prentice-Hall, Englewood Cliffs, N.J., 1965.

D. W. Mathieson (Ed.), *Interpretation of Organic Spectra.* Academic Press, New York, 1965.

R. M. Silverstein, and G. C. Bassler, *Spectrometric Identification of Organic Compounds,* 2nd Ed. John Wiley and Sons, New York, 1967.

D. H. Williams, and I. Fleming, *Spectroscopic Methods in Organic Chemistry.* McGraw-Hill, New York, 1966.

Infrared

K. Nakanishi, *Infrared Absorption Spectroscopy.* Holden-Day, San Francisco, 1962.

CHAPTER 10

NUCLEAR MAGNETIC RESONANCE SPECTROSCOPY

Protons and certain other atomic nuclei have the property of nuclear spin. When placed in a magnetic field, the spinning nucleus undergoes precession, and a change in the quantum level (spin state) can be induced by irradiation with energy in the radio-frequency region of the spectrum. The equivalence (*resonance*) of the precession frequency and irradiation frequency is a function of the strength of the magnetic field and the specific nucleus. For protons the resonance condition is usually measured at 60 megacycles per second in a magnetic field of about 14,000 gauss. The absorption of the radiation by the protons at resonance can be detected and recorded electronically.

The value of proton magnetic resonance in organic chemistry lies in the fact that the effective magnetic field around a covalently bonded proton varies slightly with its chemical environment. Therefore, when an organic compound is placed in an nmr spectrometer which can sweep an extremely narrow range of magnetic field strength or frequency, instead of one signal for all the protons, a spectrum is produced with a separate absorption for each type of proton in the molecule. The spectrum provides three basic kinds of information:

1. The signal strength, or *peak area,* as measured by electronic integration, is directly proportional to the number of identical protons in the sample which produce the signal.

2. The position of the peak in the spectrum, or *chemical shift,* is determined by the atom or structural grouping to which the proton is bonded.

3. Because of *coupling* with the magnetic fields due to spins of neighboring protons, a signal may be split into several peaks. The number (multiplicity) and separation of these peaks is characteristic of the number and steric relationship of these nearby protons.

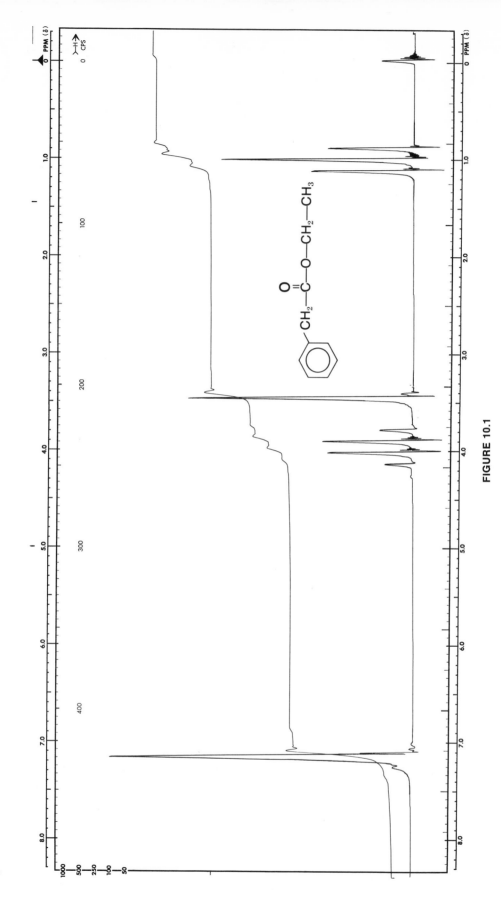

FIGURE 10.1

These three factors can be seen in the spectrum of ethyl phenylacetate (Fig. 10.1). The four main steps in the integration line, from left to right, have relative heights of 5:2:2:3, indicating the number of protons of each type. The positions of the four signals, from left to right, correspond to C_6H_5, —OCH_2—, Ph—CH_2—CO, and —CH_3 protons. Finally the splitting of the —OCH_2— and —CH_3 signals into a quartet (4 peaks) and a triplet (3 peaks) indicates that each group is adjacent to the other in the molecule.

PEAK AREAS

As previously indicated, the relative peak areas are given by the changes in height of the integration curve. These may be obtained by counting lines if the spectrum is recorded on ruled paper or by measuring with a ruler. From the measured heights, derive the smallest set of integers which are in the same ratio. Thus the integration steps in Figure 10.1 are from left to right 60, 24, 24, and 36 mm on the original spectrum. Dividing each by 12 provides the integer ratios 5:2:2:3 as mentioned. The error in electronic integration is often 5 to 10 per cent so that ratios such as 1.1:2 and 2.8:5 should be rounded off to 1:2 and 3:5, respectively. With multiplets or overlapping signals, the area of the entire grouping is taken together.

If several signals are present, it is often possible to make a fairly good estimate of the total number of protons in the compound, as was done in the foregoing. It should be recognized, however, that any multiple of a ratio may be correct; i.e., two signals in a ratio of 1:2 may correspond to a total of 3, 6, 9 . . . . protons in the compound.

CHEMICAL SHIFTS

The chemical shift of a peak (or "line") in a spectrum is related to the environment of the proton(s) causing the signal. If the proton is strongly *shielded* by electrons around it, screening it from the external magnetic field, it appears at *high field* (to the right side of the spectrum). Conversely, if adjacent atoms withdraw electrons from around the proton, it is *deshielded* and its signal appears at lower field. Because of special magnetic effects (anisotropy), protons attached to unsaturated atoms are additionally deshielded and appear at lower fields.

The chemical shift of a signal is expressed in terms of its position relative to the peak due to tetramethylsilane [TMS, $(CH_3)_4Si$]. A small amount of TMS is usually added to the sample as an internal standard when the spectrum is run. The TMS protons appear at very high field, well separated from the signals of the protons in common compounds. On the scale that we will use, the TMS protons are assigned a value of $\delta = 0$, and chemical shifts of other protons are expressed in parts per million (ppm) downfield from the TMS signal. The chemical shift can also be expressed in cycles per second (cps or Hz); for a 60 MHz instrument, the value in cps is given

TABLE 10.1 *CHEMICAL SHIFTS OF ALIPHATIC PROTONS*[a]

PROTON	δ(ppm)	PROTON	δ(ppm)
CH_4	0.23	$R'O—CH_3$	3.2–3.5
$R—CH_3$	0.8–1.2	$R''OCH_2CH_3$	1.2–1.4
R_2CH_2	1.1–1.5	$R''O(CH_2)_2CH_3$	0.9–1.1
R_3CH	1.4–1.6	$ArO—CH_3$	3.7–4.0
$R'CH{=}CH—CH_3$	1.6–1.9	$R'C(\!\!=\!\!O)OCH_3$	3.6–3.9
$Ar—CH_3$	2.2–2.5	$ArC(\!\!=\!\!O)OCH_3$	3.7–4.0
$R—C(\!\!=\!\!O)CH_3$	2.1–2.4	$R_2'N—CH_3$	2.2–2.6
$Ar—C(\!\!=\!\!O)CH_3$	2.4–2.6	$ArN(R')CH_3$	2.8–3.1
$R'O—C(\!\!=\!\!O)CH_3$	1.9–2.2	$R''C(\!\!=\!\!O)N(CH_3)R$	2.8–3.0
$ArO—C(\!\!=\!\!O)CH_3$	2.0–2.5	$I—CH_3$	2.15
$R_2''N—C(\!\!=\!\!O)CH_3$	1.8–2.2	$Br—CH_3$	2.7
$N{\equiv}C—CH_3$	2.00	$Cl—CH_3$	3.0
$O_2N—CH_3$	4.33	$F—CH_3$	4.3

[a]Compiled from spectra and other data given in references 1 to 4. In this table R = alkyl, R′ = alkyl or H, Ar = aryl, R″ = alkyl, aryl, or H.

by δ(ppm) × 60. Both scales generally appear on the chart paper used for recording nmr spectra.

Table 10.1 lists typical values of the chemical shift for aliphatic protons in various environments. Several useful generalizations can be drawn from the data in these tables:

1. The protons in methyl groups (RCH_3) appear at higher field (lower δ value, further to the right in the spectrum) than those in methylene groups (R_2CH_2); methine protons (R_3CH) appear at still lower field (see first four entries in Table 10.1). For example, the three different types of protons in 2,4-dimethylpentane have the following chemical shift values:

$$
\begin{array}{c}
\overset{1.61}{\underset{}{\text{H}}} \qquad \text{H} \\
\overset{0.88}{\text{CH}_3}\text{—}\overset{|}{\text{C}}\text{—}\overset{1.04}{\text{CH}_2}\text{—}\overset{|}{\text{C}}\text{—CH}_3 \\
\underset{\text{CH}_3}{|} \qquad \underset{\text{CH}_3}{|}
\end{array}
$$

2. For protons on carbon attached to an electronegative atom or group X, the chemical shift (δ value) increases with the electronegativity of X. This is due to the inductive effect on the shielding of the protons and is apparent in the methyl halides given at the end of Table 10.1.

3. The inductive (deshielding) effect of a substituent on a proton decreases as the separation between the proton and substituent is increased. This is illustrated by the first three entries in the second column of Table 10.1. Specific examples of this attenuation of the deshielding can be seen in the δ values of the protons in $CH_3CH_2CH_2$—X compounds:

$$CH_3\text{—}CH_2\text{—}CH_2\text{—X}$$

X = NO_2	1.03	2.07	4.38
X = OH	0.92	1.57	3.58
X = CHO	0.97	1.67	2.42

The effect of two deshielding substituents on the chemical shifts of methylene protons is cumulative, although not exactly additive. Estimates of the chemical shift in a group X—CH_2—Y can be made by adding the "effective shielding constants" $\Delta\delta$ for X and Y (Table 10.2) to 0.23, which is the chemical shift for the protons in methane. For example, the chemical shift for the CH_2 group in ethyl phenylacetate is calculated to be: $0.23 + 1.85 + 1.55 = 3.63$ ppm; the observed value (Fig. 9.1) is 3.48.

Protons on an aromatic ring appear at very low field (e.g., benzene, $\delta 7.27$), due to the aromatic "ring current." Electron withdrawing groups cause downfield shifts relative to benzene, and electron releasing groups cause upfield shifts. In addition, substituents with significant (diamagnetic)

anisotropy (e.g., —NO_2 and —$\overset{|}{C}$=O) deshield the *ortho* protons even more, often as much as 0.5 to 1.0 ppm. Representative values for three com-

TABLE 10.2 EFFECTIVE SHIELDING CONSTANTS IN X—CH$_2$—Y[a]

GROUP	Δδ	GROUP	Δδ
CH$_3$—	0.65[b]	HO—	2.56
R$_2$C=CR—	1.32	RO—	2.36
C$_6$H$_5$—	1.85	ArO—	3.23
R—C(=O)\	1.70	RC(=O)(—O—)	3.13
ROC(=O)\	1.55	R$_2$N—	1.57

[a]B. P. Dailey, and J. M. Shoolery, *J. Am. Chem. Soc.*, 77, 3977 (1955).
[b]A value of 0.47 was originally suggested[a] for the CH$_3$ group. The value for a methyl or alkyl group varies from 0.40 to 0.80, depending on the second substituent.

pounds are shown below (see also Fig. 10.3). Although the protons on an aromatic ring may not be identical, occasionally near coincidence of chemical shifts can cause the aromatic protons to appear as a singlet. Usually, however, different proton chemical shifts and the splitting which results because of coupling lead to several lines in the aromatic proton region of the spectrum.

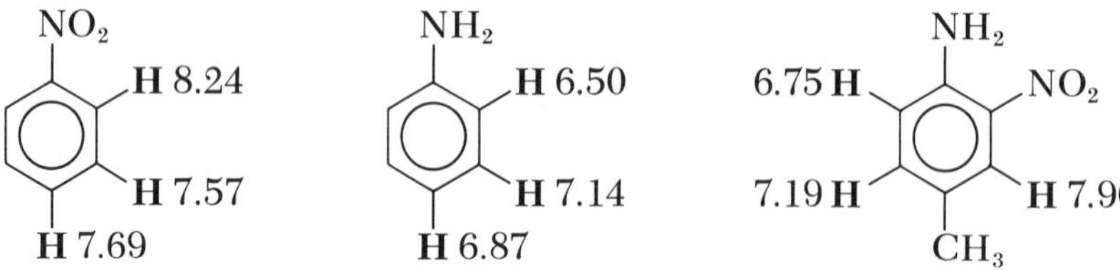

NO$_2$ H 8.24   H 7.57   H 7.69

NH$_2$ H 6.50   H 7.14   H 6.87

NH$_2$ 6.75 H   7.19 H   NO$_2$ H 7.90   CH$_3$

Inductive and anisotropic effects on shielding also effect the chemical shifts of protons on other sp^2 hybridized carbons, e.g., $\diagdown$C=C$\diagup$ and $\diagup$O=C$\diagdown$H (Table 10.3). This is particularly noticeable in the latter case (formyl protons) where the combined deshielding shifts the resonance to δ 8 to 10 ppm.

Protons attached to atoms other than carbon (i.e., O, N, S) vary widely in chemical shifts, and their position and appearance in the spectrum depend markedly upon the temperature, solvent, concentration, and the presence of acidic or basic impurities. The ranges of chemical shifts shown in Table 10.4 are for 5 to 50 per cent solutions in nonpolar solvents, i.e., the concentrations at which they are normally measured. Amide N—H protons

TABLE 10.3 *CHEMICAL SHIFTS OF PROTONS ON UNSATURATED CARBON ATOMS*[a]

PROTON	δ(ppm)	PROTON	δ(ppm)
$RCH{=}CH_2$	4.9–5.2	$R'{-}\overset{\displaystyle O}{\underset{\displaystyle H}{C}}$	9.4–10.2
$RCH{=}CH_2$	5.8–6.0	$R'O{-}\overset{\displaystyle O}{\underset{\displaystyle H}{C}}$	8.0–8.2
$R_2C{=}CH_2$	4.5–5.1	$R'_2N{-}\overset{\displaystyle O}{\underset{\displaystyle H}{C}}$	7.9–8.2
$RCH{=}CHR$	5.2–5.7	$R{-}C{\equiv}CH$	2.3–2.5[b]
$R_2C{=}CHR$	5.1–5.5	$ArC{\equiv}CH$	2.8–3.1
$ROCH{=}CH_2$	4.0–5.0	$RC\overset{\displaystyle O}{\underset{\displaystyle CH=CHR'}{}}$	5.7–6.2
$ROCH{=}CH_2$	6.0–7.5	$RC\overset{\displaystyle O}{\underset{\displaystyle CH=CHR'}{}}$	6.8–7.2

[a]Cf. footnote Table 10.1.   [b]Exception: Propyne δ 1.80 ppm.

often appear as broad signals which may be difficult to recognize. Amines and alcohols usually give **OH** and **NH** proton signals which are singlets regardless of the number of adjacent protons. This is caused by chemical exchange of protons which occurs at a rate greater than the frequency (observation time) of the nmr instrument. Due to this exchange, protons on N, O, and S can be replaced by deuterium by shaking the sample with D_2O. The result is the disappearance of the signal for these protons from the spectrum and the appearance of a signal for **HOD**.

TABLE 10.4 *CHEMICAL SHIFTS OF PROTONS BOUND TO O, N, S*

PROTON	δ(ppm)	PROTON	δ(ppm)
ROH	3–6	RNH_2	0–2.5
ArOH	6–8	$ArNH_2$	3–4.5
R'COOH	10–12	$R'CONH_2$	5.5–7.5
		R—SH	1–2
(enol, 14–16)	14–16	Ar—SH	3–4

COUPLING CONSTANTS

The most frequently encountered spin coupling is that between protons on adjacent saturated carbon atoms, as seen in the ethyl protons of Figure 10.1. In simple systems of this type, the signal of a proton which has n identical adjacent protons is split into $n+1$ peaks (lines). Thus, in Figure 10.1, the OCH_2 protons are adjacent to three identical methyl protons and appear as a quartet (4 lines). Similarly the CH_3 proton signal appears as a triplet due to coupling with two adjacent methylene protons. The spacing of the lines (splitting) in the triplet is exactly equal to that in the quartet, and in simple spectra, this spacing is equal to a theoretical quantity, the coupling constant, J. Typical values of coupling constants between variously related pairs of protons are given in Table 10.5.

TABLE 10–5 *REPRESENTATIVE VALUES OF COUPLING CONSTANTS*

When several nonidentical protons are mutually coupl[ed], very complex multiplets can result, as illustrated by the spectrum of n-[bu]tyl bromide (Fig. 10.2). However, even in this case the peaks can be s[een] to fall into three groups at δ3.7–3.4, 2.2–1.2, and 1.2–0.7 ppm. The in[teg]ration curve suggests a ratio of 2:4:3 protons in these regions, respect[ive]ly, and only the relatively deshielded —CH_2Br protons are easily re[cog]nizable as a triplet, i.e., coupled to an adjacent —CH_2— group. The c[hem]ical shift of these protons (δ 3.53) is measured at the center of the tr[ipl]et. The high field grouping of peaks (3 protons) is from the —CH_3 p[roto]ns, and this distorted triplet is typical of longer chain alkyl groups. The [cen]tral multiplet of four protons is uninterpretable due to the extensive coupling and similarity of chemical shifts.

The aromatic ring protons of 4-methyl-3-nitroaniline are also mutually coupled as shown in Figure 10.3. In this case the coupling aids in interpreting the spectrum rather than complicating it. By comparison of the total integration shown with that of the methyl group (at δ 2.42, not shown), it is found that there are 3 protons on the ring. The first proton (A) appears as a doublet ($J = 2$ Hz) at 7.23 ppm, the second (B) as a doublet ($J = 7$ Hz) at 7.05 ppm, and the third (C) as a doublet of doublets ($J = 2, 7$ Hz) at 6.75 ppm. The magnitudes of the coupling constants (see Table 10.5) indicate that **H**$_A$ is *meta* to **H**$_C$, and **H**$_B$ is *ortho* to **H**$_C$; the coupling constant between the *para* protons, **H**$_B$ and **H**$_C$, is approximately zero. The substituents are therefore in the 1,2, and 4 positions of the benzene ring. **H**$_A$ appears at lowest field because it is *ortho* to the deshielding —NO$_2$ group, and **H**$_C$ at highest field since it is *ortho* to the NH$_2$ group (see p. 87).

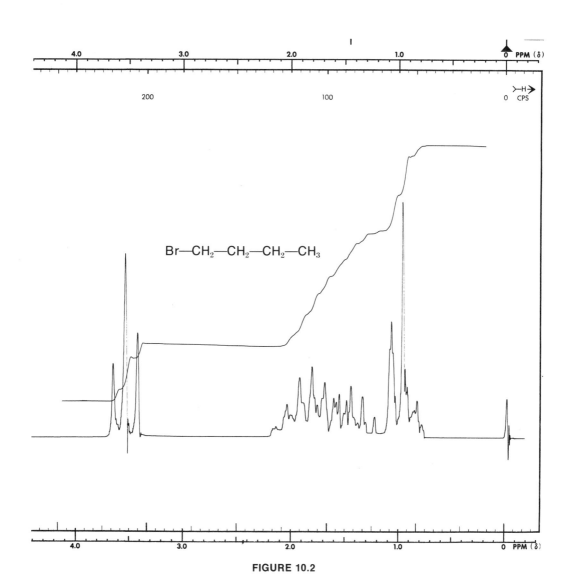

FIGURE 10.2

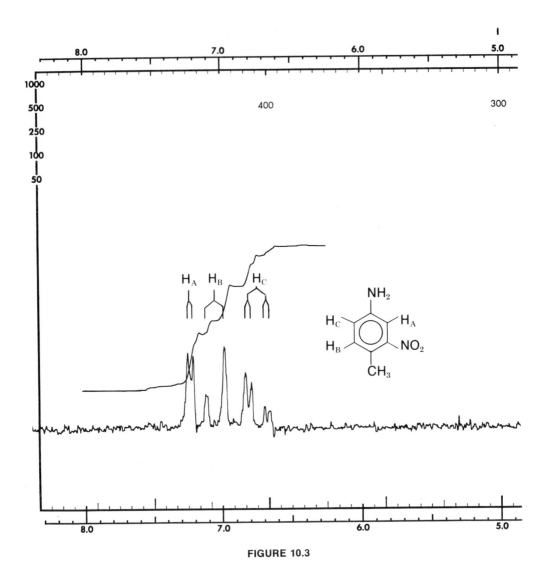

FIGURE 10.3

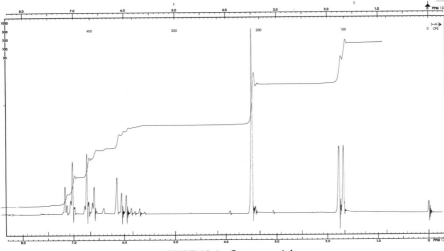

FIGURE 10.4   Compound A.

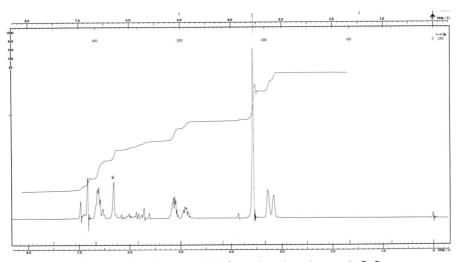

FIGURE 10.5   Compound B. Starred peak exchanges in D$_2$O.

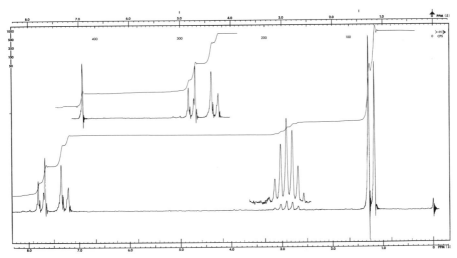

FIGURE 10.6   Compound C. Upper trace offset 3 ppm.

QUESTION

In Figures 10.4 to 10.6 are given the nmr spectra of compounds A, B, and C (see question at end of Chapt. 9). Identify the compounds, and assign and explain the peaks in the spectra.

References

J. D. Roberts, *Nuclear Magnetic Resonance.* McGraw-Hill, New York, 1959. (See also references in Chapter 9.)

Varian Associates, *High Resolution NMR Spectra Catalog.* Palo Alto, Calif., 1962–1963. (In two volumes.)

D. W. Mathieson (Ed.), *Nuclear Magnetic Resonance for Organic Chemists.* Academic Press, New York, 1967.

K. Nakanishi, V. Woods, and L. J. Durham, *A Guidebook to the Interpretation of NMR Spectra.* Holden-Day, San Francisco, 1967.

CHAPTER 11

THE LITERATURE OF ORGANIC CHEMISTRY

Since the beginnings of chemistry, a continuous record of experimental fact and interpretation has been evolving into what we call the chemical literature. One of the major elements in this body of knowledge is the description of the individual compounds that have been encountered and characterized by chemists in the past 150 years.

The student of organic chemistry quickly becomes aware that a very large number of organic compounds can exist, and he learns to write structures and devise synthetic routes without particular concern as to whether the compound in question has actually been prepared or isolated and described somewhere in the literature. This point becomes of crucial importance, however, if one needs a sample of the compound or information on some property. The problem of finding a specific compound in the literature and locating information on preparative routes, physical constants, or reactions is a practical and exceedingly important one. An introduction to this type of literature search is given in this chapter.

Organic chemists have been confronted for over a century with the problem of keeping track of compounds and of coordinating data on new as well as previously known compounds in a systematic way. The organization of literature dealing with organic compounds has become an increasingly complex task since their number grows exponentially. The sheer bulk of this material is difficult to comprehend; the total number of known compounds is currently estimated to be over 4,000,000.

Ultimately, the only possibility for coping with this mass of material in a systematic way is the use of computer techniques for storing and retrieving information. A program to record all new compounds appearing in the current literature, and eventually all compounds that have been described in the past, is underway at the Chemical Abstracts Service of the American Chemical Society. This computer registry includes data on over a million compounds, but these data are from relatively recent years only. The computer registry will permit all-embracing searches based upon structural formulas and also substructural units, something that could be done in no other way. On the other hand, it is quite certain that existing sources and methods of searching for information on compounds will be in use for a number of years to come.

SEARCHING THE LITERATURE FOR COMPOUNDS

The primary source of chemical information of any kind is the original report of the data in a journal or patent. In order to make this information more accessible, numerous encyclopedias, indexes, and reviews have been and are continuing to be published. Several of the secondary sources which are useful in locating syntheses and properties of organic compounds are described here. An excellent summary of literature sources has been provided by J. E. H. Hancock, *J. Chem. Educ.*, **45**, 193, 260, 336 (1968).

Handbooks.   For quick reference to physical properties (melting and boiling points, solubility, density) of simple organic compounds, several books of tables are available. In most of these, references are given to the source of the data. The following represent a few of the most common and useful references.

 a. Physical Constants of Organic Compounds, In *Handbook of Chemistry and Physics*, 50th Ed. Chemical Rubber Co., Cleveland, 1969. Contains mp, bp, solubility, and other data for approximately 14,000 organic compounds.
 b. Physical Constants of Organic Compounds, In N. A. Lange (Ed.), *Handbook of Chemistry*. McGraw-Hill, New York. Similar to the preceding handbook but less extensive in coverage (approximately 7000 compounds in the 10th edition, 1961).
 c. *Handbook of Tables for Identification of Organic Compounds*, 3rd Ed. Chemical Rubber Co., Cleveland, 1967. Contains mp and bp data for over 4000 compounds and melting points of derivative compounds for each. Arranged by functional groups.

Other Reference Works

Chemistry of Carbon Compounds, E. H. Rodd (Ed.) Vols. I (1951)–V(1962). Elsevier, N.Y.; *Rodd's Chemistry of Carbon Compounds*, S. Coffey (Ed.). Elsevier New York, 1964. A multivolume survey of all classes of organic compounds, with properties and preparative methods given for many typical compounds. Second edition still in the process of being published.

Dictionary of Organic Compounds (Heilbron), 4th Ed., Vols. 1–5 (1965) and annual supplements, Oxford University Press, New York. The Heilbron Dictionary is an alphabetical listing of some 25,000 compounds with selected reactions, derivatives, and literature references.

Beilstein Handbuch der Organische Chemie, Springer-Verlag, Berlin, 1918 to the present. This German treatise is the ultimate secondary source on all organic compounds in the earlier literature (prior to 1940). The fourth edition, in 31 volumes (*Bände*) was published in the period 1918–1938, and this *Hauptwerk* (HW) covered exhaustively the literature through the year 1909. Two further complete coverages, with parallel organization in each volume, were compiled as supplements. The first (*Erstes Ergänzungswerk*, E-I, 1928–1938) covers the period 1910–1919; the second (*Zweites Ergänzungswerk*, E-II, 1941–1957), the period 1920–1929. A third supplement is now being issued (E-III, 1958 to the present), covering the period 1930–1939, with many more recent references also included.

With a rudimentary understanding of German nomenclature it is possible to locate a compound in Beilstein using the formula index (*General Formelregister*) for the second supplement. This index also gives references to entries for the same compound in the HW and E-I series. Alternatively, one can make use of the organization of the volumes to seek out the compound directly. To do the latter some understanding of the classification system is necessary.

All of the compounds in Beilstein are classified in one or another of three main categories: Aliphatic (*Acyclische*), Carbocyclic (*Isocyclische*, including both alicyclic and aromatic rings), and Heterocyclic (*Heterocyclische*). Heterocyclic compounds are further subdivided according to the type and number of heteroatoms in the ring(s). Thus *Band XX* (Volume 20), *Heterocyclische Verbindungen, Heteroclasse 2 O bis 9 O*, contains heterocyclic compounds with from two to nine oxygen atoms in the ring(s).

Within each of the main series, compounds are arranged according to the functional group present. The major functional group categories are *Kohlenwasserstoffe* (hydrocarbons), *Oxyverbindungen* (alcohols), *Oxoverbindungen* (aldehydes and ketones), *Carbonsäuren* (carboxylic acids), *Sulfinsäuren* (sulfinic acids), *Sulfonsäuren* (sulfonic acids), *Amine, Hydroxylamine, Hydrazine, Azoverbindungen,* and *Phosphine*. The *Stammkerne* sections in the *Heterocyclische Bande* contain heterocyclic compounds with none of the foregoing functional groups; i.e., they are analogous to the *Kohlenwasserstoffe* sections of the *Acyclische* and *Isocyclische* categories. In addition to the preceding functional group categories, there are separate sections for compounds containing more than one major functional group. The titles of these are self-explanatory: *polycarbonsäuren, oxocarbonsäuren, oxyoxoverbindungen, oxoamine,* and so forth.

Compounds containing halogen atoms, nitro, nitroso, or azide groups are listed under the parent compound; e.g., chloroacetone under acetone. Esters, primary amides, anhydrides, nitriles, acid halides are listed under the parent carboxylic acids. Secondary and tertiary amides are given under the parent amine. Ethers, peroxides, mercaptans, sulfides, and sulfones are given following the corresponding alcohols.

Within any functional group category, the compounds are ordered according to the number of oxygen, sulfur, or nitrogen atoms contained. These subgroups are further ordered according to the degree of unsaturation (number of rings and double bonds) and finally according to the number of carbon atoms present. Isomeric compounds are generally arranged so that the simpler (less branched, fewer rings, and so forth) isomers precede the more complex. The preceding classification is outlined in the Table of Contents (*Inhalt*) at the beginning of each volume.

As an example of how to use the Beilstein system of organization, let us see how to find methyl 5-chloro-2-hydroxybenzoate (methyl 5-chlorosalicylate).

The compound is first recognized as a carbocyclic (*Isocyclische*) hydroxy-acid (*Oxycarbonsäuren*) derivative, and should be found in *Band X*, which is so titled. The parent compound, salicylic acid, has a molecular formula of $C_7H_6O_3$, i.e., $C_nH_{2n-8}O_3$ (n = 7). Looking in the Table of Contents of *Band X*, we first locate the subheading *oxycarbonsäuren mit 3 sauerstoffatomen* (hydroxyacids with 3 oxygen atoms). In this section we then look for compounds with five units of unsaturation (i.e., $C_nH_{2n-8}O_3$), and finally the subheading for n = 7. At this point we are directed to *Seite*, (page) 43.

Turning to page 43, we find salicylic acid itself; in fact pages 43 to 59 discuss salicylic acid. Continuing on we find salts of salicylic acid, ethers of salicylic acid, esters of salicylic acid, and so forth. On page 101 begins the listing of substituted salicylic acids, the first compounds being chloro-salicylic acids. After the 5-chloro isomer is a paragraph on page 103 describing its methyl ester. This is shown in Figure 11.1. Similarly a supplemental entry for this compound can be found in E-II, *Band X*, *seite* 62 (Fig. 11.2). There is no information on this compound in E-I, and *Band X* of E-III has not yet been published.

Both of the preceding entries can be located equally well using the *General Formelregister* of E-II. Of the 32 isomers given under $C_8H_7ClO_3$, the compound desired is listed as *5-Chlor-salicylsäure-methylester* (Fig. 11.3).

Chemical Abstracts. Publication of *Chemical Abstracts* was begun in 1907, and since 1945 the volume and collective indexes have become the "key to the chemical literature." All new compounds appearing in chemical journals (nearly 12,000 periodicals are now abstracted) and patents are recorded in these indexes, as well as entries on most citations of previously known compounds. The most recent complete collective index (1962–1966) contains 8 million entries and occupies 40,000 pages. Despite this huge bulk, a compound mentioned in any abstract published in this 5-year period can be found in a few minutes time, after a little practice, and traced back from index to abstract and thence to the original primary journal description. Ten-year collective indexes were published for the first five decades; the collective indexes are now on a five-year basis. Only these collective indexes need to be consulted for the period prior to 1967, but individual volume indexes (now semiannual) must be consulted until the eighth collective index is published.

5-Chlor-salicylsäure-methylester $C_8H_7O_3Cl = HO \cdot C_6H_3Cl \cdot CO_2 \cdot CH_3$. *B*. Beim Ein-leiten von Chlorwasserstoff in eine methylalkoholische Lösung der 5-Chlor-salicylsäure (VARN-HOLT, *J. pr.* [2] **36**, 21). Bei der Einw. von Methyljodid auf das Silbersalz der 5-Chlor-salicyl-säure (LASSAR-COHN, SCHULTZE, *B.* **38**, 3300). — Nadeln (aus Alkohol). F: 48° (SMITH, *B.* **11**, 1227; V.), 50° (L.-C., SCH.). Siedet unter teilweiser Zersetzung bei 249° (V.). Ziemlich leicht löslich in Alkohol (SM.; V.).

FIGURE 11.1 *Beilstein, Hauptwerk,* Band X, p. 103.

5-Chlor-salicylsäure-methylester $C_8H_7O_3Cl = HO \cdot C_6H_3Cl \cdot CO_2 \cdot CH_3$ (H 103). *B*. Bei der Belichtung einer Lösung von Chlorpikrin in Salicylsäuremethylester (PIUTTI, BADOLATO, *R. A. L.* [5] **33 I**, 477). — F: 48°.

FIGURE 11.2 *Beilstein,* E-II, *Band X*, p. 62.

$C_8H_7ClO_3$ Chlormethyl-phenyl-carbonat **6** I 88.
2-Chlor-phenoxyessigsäure **6** II 172.
4-Chlor-phenoxyessigsäure **6**, 187, II 177.
Brenzcatechin-monochloracetat **6** II 783.
Chlorameisensäure-[2-methoxy-phenylester]
    6, 776, I 386.
Chlorhydrochinon-acetat **6**, 849, I 417.
5-Chlor-4-oxy-3-methoxy-benzaldehyd
    8 II 286.
ω-Chlor-2.4-dioxy-acetophenon **8** I 615,
    II 296.
ω-Chlor-3.4-dioxy-acetophenon **8**, 273, I 618,
    II 299.
3-Chlor-salicylsäure-methylester **10**, 101,
    II 61.
Methyläther-5-chlor-salicylsäure **10**, 103,
    II 62.
5-Chlor-salicylsäure-methylester **10**, 103,
    II 62.
Methyläther-6-chlor-salicylsäure **10**, 104.
2-Chlor-3-oxy-benzoesäure-methylester
    10, 142.
3-Chlor-2-methoxy-benzoesäure **10** II 61.
4-Chlor-2-methoxy-benzoesäure **10** II 62.
2-Chlor-3-methoxy-benzoesäure **10** II 83.
4-Chlor-3-methoxy-benzoesäure **10** II 83.
6-Chlor-3-methoxy-benzoesäure **10**, 143,
    II 83.
6-Chlor-3-oxy-benzoesäure-methylester
    10, 143.
2-Chlor-4-methoxy-benzoesäure **10**, 175.
3-Chlor-4-methoxy-benzoesäure **10**, 176.
3-Chlor-4-oxy-benzoesäure-methylester
    10, 176, II 102.
5(?)-Chlor-2-oxy-phenylessigsäure **10**, 189.
2-Chlor-mandelsäure **10** II 124.
4-Chlor-mandelsäure **10**, 210, I 92.
5-Chlor-2-oxy-3-methyl-benzoesäure
    10 II 133.
4-Oxy-3-chlormethyl-benzoesäure **10**, 226.
6-Oxy-3-chlormethyl-benzoesäure **10**, 231.
5-Chlor-2-oxy-4-methyl-benzoesäure
    10, 236, I 101, II 140.
3-Oxy-x-chlormethyl-benzoesäure **10**, 241.
Piperonal-hydrochlorid **19**, 120.

FIGURE 11.3 *Beilstein,* Formula Index, E-II, p. 405.

Compounds are indexed both by empirical formula and by the chemical name. If a specific compound is wanted, the formula index is generally consulted first, and after the compound in question is located by name out of a list of perhaps ten or twenty isomeric compounds, an abstract column or number and also the subject index name are obtained. This procedure may immediately turn up references to the desired compound, but it usually pays to go from the formula index to the subject index, using the name obtained in the former. The subject index may lead to additional entries, and some indication of the information available in the reference will be given.

It should also be mentioned that in the early volumes of *Chemical Abstracts*, the formula indexes did not provide as complete a coverage as the subject index.

The main value of the subject index is its correlative function. The system used for index names is designed to group closely related compounds together; thus a series of esters of a complex acid will be found under the main entry for the acid; e.g., *Salicylic acid, 5-chloro-, methyl ester,* is the subject index "inversion name," and other derivatives of the carboxyl group will be listed similarly under this heading. Some experience is required to find a compound in the subject index; a detailed discussion of the nomenclature system is given in a preface to the index of Vol. 56 (1962). For this reason, except for simple compounds, the formula index provides a surer way to find a specific compound, but if any of several related compounds would be equally useful, each must be searched for individually in the formula index. The *Chemical Abstracts* references to compound I, above, are shown in Figures 11.4 to 11.6.

SEARCHING FOR PREPARATIVE METHODS

When the preparation of a compound is required, the first step is to locate the substance in the literature. All sources or preparations of a compound recorded in the time period covered by a volume of Beilstein will be cited there. A very convenient compilation of synthetic methods that have been used for relatively simple compounds up to 1950 is *Synthetic Organic Chemistry* by Wagner and Zook, John Wiley, New York, 1953. The only way to check systematically for more recent syntheses is a search of *Chemical Abstracts* subject indexes. This entails some labor since a number of entries may be found, and not all of those pertaining to synthesis will necessarily be identified as such. Some entries may refer to uses, analytical methods, and the like, and can be ignored; abstracts corresponding to the other entries must then be scanned.

The method or methods described for the compound must then be evaluated from the standpoint of two primary criteria: (1) simplicity and practicality, and (2) economy in terms of yield from readily available starting materials. Under the first heading come such items as need for special

C$_8$H$_7$ClO$_3$ Acetic acid, (chlorophenoxy)-, **16**:
3483[7]; **21**:1096[7]; **24**:1344[7]; **25**:930[2];
30:2963[5]; **36**:6199[7], 6512[5]; **37**:4426[2],
5637[4]; **38**:2327[6]; **39**:1199[4], 5286[8];
40:1474[1], P 2264[5], 3846[8], 6197[4].[7]; *Na
salt*, **40**:7491[6].

Acetophenone, chlorodihydroxy-, **18**:
675[3]; **23**:2161[3]; **24**:4012[9]; **29**:3338[1],
5839[5]; **30**:159[9], 3421[9]; **34**:401[6]; **37**:
101[3], 2358[2]; **39**:1471[4].

Anisic acid, 3-chloro-, **38**:3629[9]; **40**:2131[1].

Benzoic acid, 3-chloro-4-hydroxy-, methyl
ester, **20**:3712[8].

—, chloromethoxy-, **18**:386[6]; **20**:1065[3];
21:3189[8]; **23**:1128[9]; **24**:1859[2]; **34**:
7867[4]; **35**:2125[6]; **38**:5495[7]; *and salts*,
33:2124[6].[7].

Carbonic acid, chloromethyl phenyl ester,
14:739[9].

Cresotic acid, chloro-, **25**: P 4558[6]; **26**: P
1130[5], P 1946[1], P 2604[3]; **27**: P 852[5], P
2044[1]; **33**:6897[6].

4-Cyclohexene-1,2-dicarboxylic anhydride, 1-chloro-, **40**: P 3136[5].

2-Furoic acid, 2-chloroallyl ester, **32**: P
7925[5].

Homoprotocatechuyl chloride, **32**:851[7].

Mandelic acid, chloro-, **15**:2632[2]; **22**:
2746[4]; **25**:3637[4]; **32**:4969[2]; **35**:5637[2];
40:1156[6], 3044[3].

p-Orsellinaldehyde, 3-chloro-, **28**:6130[5];
29:1078[7].

Piperonyl alcohol, 6-chloro-, **33**:1294[1].

Salicylic acid, chloro-, Me ester, **31**:
8599[6]; **33**:2124[5]; **37**:2010[3]; **38**:3256[1].

Sorbic acid, γ-chloroacetyl-δ-hydroxy-,
δ-lactone, **35**:7405[1].

Toluic acid, chlorohydroxy-, **25**:3325[5];
37:3420[7].

Vanillin, chloro-, **19**:2494[8]; **20**:1980[6];
21:906[1]; **23**:4456[5]; **25**:94[6].[7]; **40**:3896[6].

FIGURE 11.4 *Chemical Abstracts*, Formula Index.

Salicylic acid

droxyphenylazo) - **3** - hydroxy - **4**-
biphenylylazo] - **5** - hydroxy - **3**-
methyl - 1 - pyrazolyl|phenylazo]-
5 - hydroxy - **3** - methyl - 1 - pyrazo-
lyl]-, **40**: P 6265[3].

——, chloro-, effect on dermatophytes, **34**:
5873[2].

——, 3-chloro-, **40**:5423[4].

——, 3(and 5)-chloro-, and derivs., **33**:
2124[3].

——, 4(and 5)-chloro-, **40**:71[3].

——, 5-chloro-, **31**:8117[7]; **33**:8181[7].

4-ethoxybutyl ester, **35**: P 3738[6].

methyl ester, **37**:2010[3]; **38**:3256[1].

and methyl ester, **31**:8599[6].

methyl ester, *N*-(*p*-methoxyphenyl)benz-
imidate, **32**:1666[7].

p-nitrophenyl ester, **33**:1296[3].

——, 4-*p*-chlorobenzoyl-, **35**: P 4218[5];
36: P 2732[1].

FIGURE 11.5 *Chemical Abstracts*,
Subject Index.

N,N-Dichlorocarbamates; chlorination reactions. J.
Bougault and P. Chabrier. *Compt. rend.* 213, 400–2
(1941); cf. *C. A.* 37, 87[7].—PhOH is transformed by a
small excess of NCl$_2$CO$_2$Me (I) in AcOH into 2,4,6-
Cl$_3$C$_6$H$_2$OH, and *o*-HOC$_6$H$_4$CO$_2$Me into its 5-Cl compd.
I and (ClCH$_2$CH$_2$)$_2$S in C$_6$H$_6$ give unstable *bis(1,2-di-
chloroethyl) sulfide,* b$_{15}$° which readily yields HCl and
CHCl:CHSCHClCH$_2$Cl. Carbazole and a small excess
of I in AcOH yield *tetrachlorocarbazole*, m. 213°. BzNH$_2$
and 1 in aq. suspension afford BzNHCl, and PhCH$_2$-
CONH$_2$ gives *N-chlorophenylacetamide*, m. 120°. In
alk. soln. *2,4-dichloro-3,5-diketo-6-benzyl-*, m. 119°, and
-6-phenylethyl-, m. 130°, *tetrahydro-1,2,4-triazine* are
obtained from the Cl-free parents. *2-Chloro-3,5-diketo-
dibenzyltetrahydro-1,2,4-triazine* m. 153°. I and diphenyl-
hydantoin in alk. soln. afford *1,3-dichloro-5,5-diphenyl-
hydantoin*, m. 166° B. C. P. A.

FIGURE 11.6 *Chemical Abstracts,*
37, 2010[3] (1942).

apparatus, high pressures or temperatures, and ease of isolating the product. These requirements may not be completely satisfied for any of several reasons. The literature reference may deal with a catalytic process suitable for industrial equipment but impractical for laboratory work. On the other hand, the previous source may not have been a deliberate synthesis. The compound may have been obtained from a more complex molecule or as a minor product in the investigation of a reaction for some other purpose. A practical synthetic method requires that the compound be obtained in useful yield from more readily available or cheaper precursors. Ultimately, of course, the synthesis must go back to commercially available chemicals.

An important consideration, particularly when the literature references are not contemporary, is the possibility that the aforementioned criteria could be better met by a method not previously described for the compound. Improved synthetic procedures and reagents are constantly being developed. For example, the preparation of a primary alcohol may have been carried out prior to 1948 by reduction of a carboxylic acid ester with sodium in ethanol. Lithium aluminum hydride has completely displaced this method for laboratory work.

A number of monographs and compendia deal with preparative reactions, and reference to these sources may be necessary even if a prior preparation is to be repeated, since sufficient detail may not be given in the original paper. In applying a reaction to a compound for which it has not previously been used, general textbook knowledge is seldom sufficient, since reaction conditions and isolation procedures are needed. Some of the more important sources are:

1. Houben-Weyl *Methoden der Organische Chemie*, 4th Ed. E. Muller, (Ed.). G. Thieme Verlag, Stuttgart, 1952–

This is an encyclopedic multivolume series, in German, dealing with methods of general laboratory practice and procedures, and also with the preparation and reactions of classes of compounds. The 4th edition is incomplete, but it provides comprehensive coverage for certain major functional group classes.

2. *Organic Reactions*, John Wiley and Sons, New York, 1942–

In this series (currently 17 volumes), over 100 general reactions of preparative utility are discussed in detail, with typical experimental procedures and extensive tables of examples with references. Cumulative chapter indexes in the most recent volumes can be scanned for a desired reaction.

3. *Newer Methods of Preparative Organic Chemistry (Neuere Methoden)*, W. Foerst (Ed.). Interscience and Academic Press, New York and Verlag Chemie, Weinheim/Bergstrasse, 1948–

This series, in English and German, is a collection of review articles from the journal, *Angewandte Chemie*, treating about 60 reactions in all. The coverage in some of the articles is similar to that in *Organic Reactions*.

4. Theilheimer, *Synthetic Methods of Organic Chemistry*. S. Karger, Basel and Interscience, New York, 1948–

The annual volumes in this series contain brief descriptions of useful synthetic transformations taken from the literature of the year covered. The reactions are indexed according to a special system based on the type of bond formed.

5. *Organic Syntheses.* John Wiley and Sons, New York, 1920–

The annual volumes and four collective volumes of *Organic Syntheses* contain detailed procedures for the preparation of over 1000 compounds. Apparatus, conditions and work-up procedures are specified, and many of the syntheses illustrate general methods which can be applied to related compounds.

6. In addition to these compendia, several monographs have been devoted to certain major reactions such as Grignard condensations, reductions with complex hydrides and Friedel-Crafts reactions. Comprehensive tables of individual compounds, with literature references, are given in all of these treatises:

M. Kharasch, and O. Reinmuth, *Grignard Reactions of Non-Metallic Substances.* Prentice-Hall, Englewood Cliffs, N.J., 1954.

N. Gaylord, *Reduction with Complex Metal Hydrides.* Interscience, New York, 1956.

G. Olah, (Ed.), *Friedel-Crafts and Related Reactions.* Wiley-Interscience, New York, 1963–1965.

To illustrate the process, we can examine methods for preparing methyl 5-chlorosalicylate. The compound is not listed in Wagner and Zook. Two methods are given in Beilstein (Figs. 11.1 and 11.2). The earlier entry (Fig. 11.1) cites the preparation by HCl-catalyzed esterification of 5-chlorosalicylic acid in methanol, and gives the recrystallization solvent (needles from ethanol) and melting point (48°). The original reference,

however, to series 2 of *Journal für praktische Chemie,* is not easily accessible. In the more recent Beilstein reference (Fig. 11.2), a preparation is indicated by irradiation of a solution of trichloronitromethane (chloropicrin) in methyl salicylate:

One of the *Chemical Abstracts* citations (Fig. 11.6) indicates a preparation by chlorination of the hydroxyester with N,N-dichlorocarbamate esters. The other abstracts cited in the subject index (Fig. 11.5) lead to papers dealing with the bactericidal properties of chlorophenol derivatives.

At this point we can assess the possibilities. Without consulting the original references (which appear in Italian and French journals), it is apparent that the two chlorination procedures require rather special reagents, neither of which is very commonly used. Further checking on these reagents in contemporary sources reveals that a chlorination step is required for their preparation. On the other hand, the main starting material is a very cheap compound, methyl salicylate (oil of Wintergreen).

The alternative is esterification of 5-chlorosalicylic acid. This acid, but not the ester, is listed in chemical supply catalogs. The cost is about eight times that of methyl salicylate. The other reagents, however, are common laboratory chemicals, and the esterification is a standard operation (a procedure using sulfuric acid is described in Chapt. 23). For preparation of a small sample of the ester, this method would probably be the best choice.

For specific information on the procedure for carrying out the esterification and isolation of the product, many examples of acid-catalyzed esterifications can be found in the indexes of Organic Syntheses Collective Volumes. Another useful step at this point is to consult the section on acid-catalyzed esterification in Wagner and Zook. References to esterification of closely similar hydroxybenzoic acids will be given, and from these, a procedure in a conveniently accessible journal can usually be found.

LITERATURE PROBLEM

As an exercise in using the literature, a practical laboratory synthesis is to be found for one of the compounds in the list below. All of these compounds can be prepared in one or more steps from simple starting materials and without special apparatus. Obtain an assignment of one of the compounds from your instructor, consult the literature, and write up a complete experimental procedure for the synthesis on a scale to provide 5 to 10 g of the final product. Your report should include all relevant literature references; the experimental descriptions should be sufficiently detailed that the preparation could bē carried out without reference to the original literature.

At your instructor's direction, this preparation can then be carried out in the laboratory as a supplemental experiment. The starting materials for the synthesis should be reasonably priced commercial chemicals. Your instructor will determine at what point in the sequence your synthesis should begin; this will depend on the cost of various precursors and the amount of laboratory time available.

Compounds

1. $CH_3(CH_2)_3N(CH_3)_2$

2. $(CH_3)_3CCH_2NH_2$

3. $CH_3CH_2CHNH_2(CH_2)_3CH_3$

4. $(CH_3)_2C—CH_2COCH_3$
$\quad\quad\;\; |$
$\quad\quad NHCOC_6H_5$

Compounds (Continued)

5. $CH_3CH_2CH=CHCO_2C_2H_5$

6. $C_6H_5CH=C(CH_3)CH_2OH$

7. 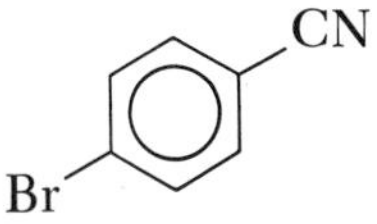 [structure: 4-bromobenzonitrile with Br and CN substituents]

8. 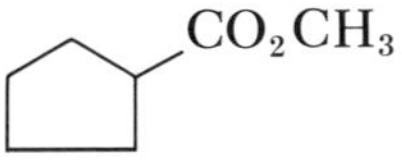 [structure: cyclopentane with CO_2CH_3]

9. [structure: 4-bromophenyl with $CH=CHCO_2H$, Br]

10. [structure: cyclohexanone with $CONHC_6H_5$, O]

11. [structure: 1-tetralone, O]

12. [structure: naphthalene with $CH=CHCO_2H$]

13. [structure: Br-coumarin with $CO_2C_2H_5$, O, O]

14. [structure: 9,10-diphenylanthracene with C_6H_5, C_6H_5]

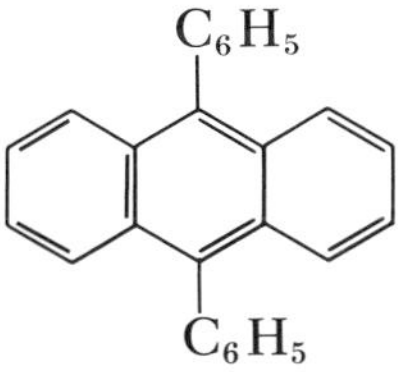

15. 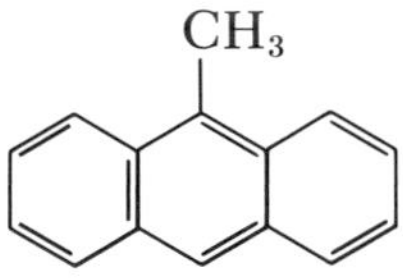 [structure: 9-methylanthracene with CH_3]

16. 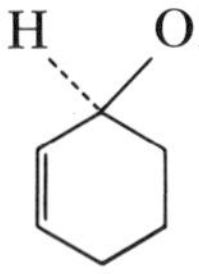 [structure: cyclohexenol with H, OH, $(S(—))$]

17. 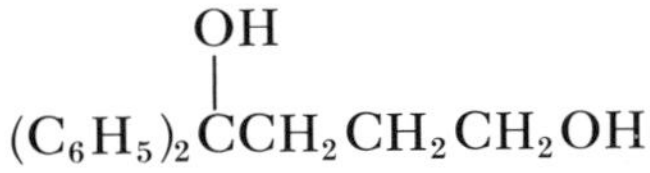 $(C_6H_5)_2\overset{\text{OH}}{C}CH_2CH_2CH_2OH$

18. 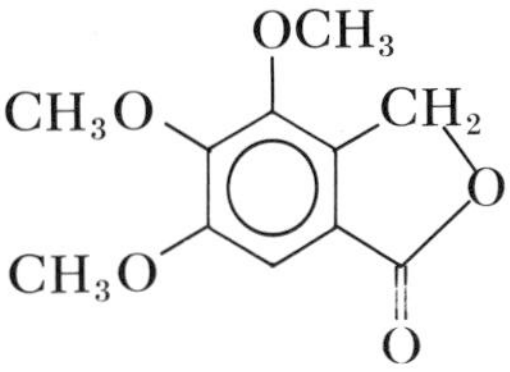 [structure: with OCH_3, CH_3O, CH_3O, CH_2, O, O]

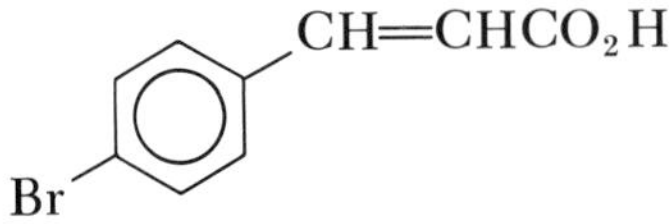

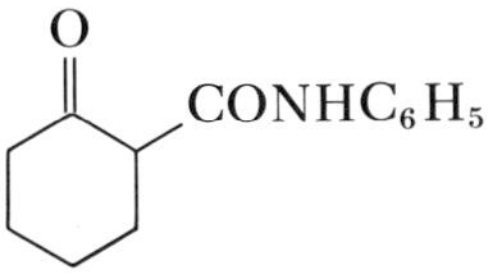

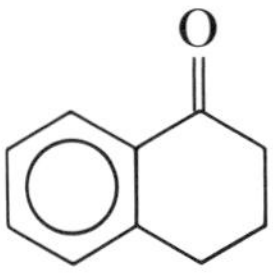

19. 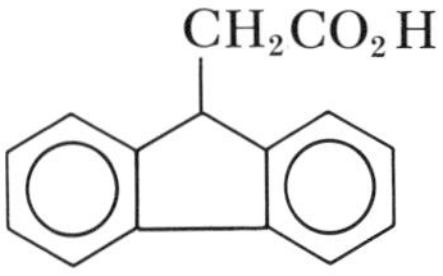 [structure: fluorene with CH_2CO_2H]

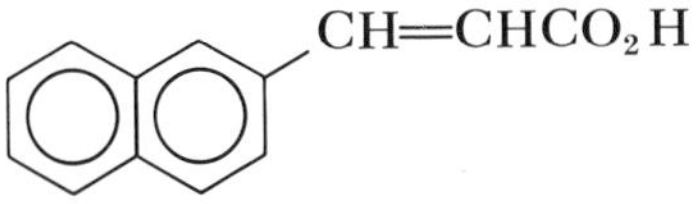

20. 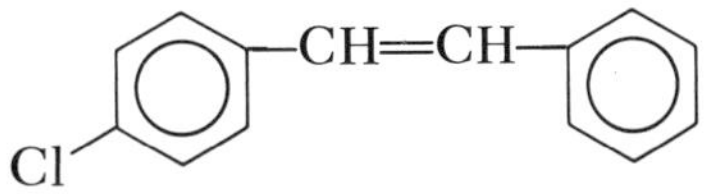 [structure: with Cl, $CH=CH$ diphenyl]

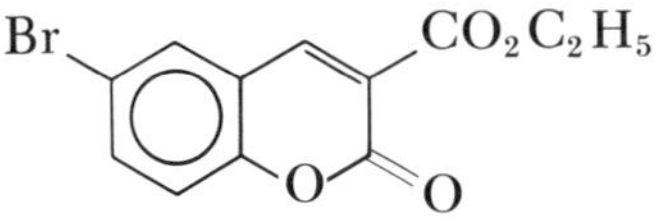

21. 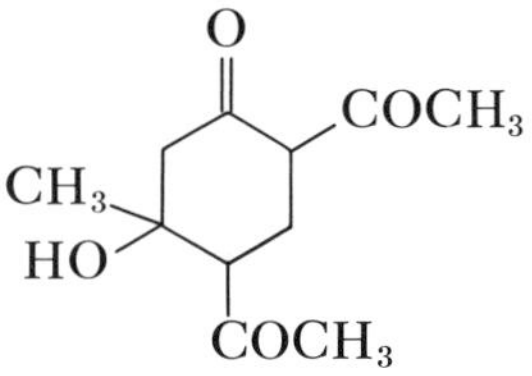 [structure: cyclohexanone with O, $COCH_3$, CH_3, HO, $COCH_3$]

22. 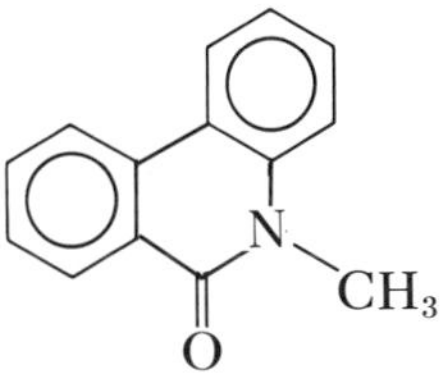 [structure: phenanthridinone with N, CH_3, O]

23. 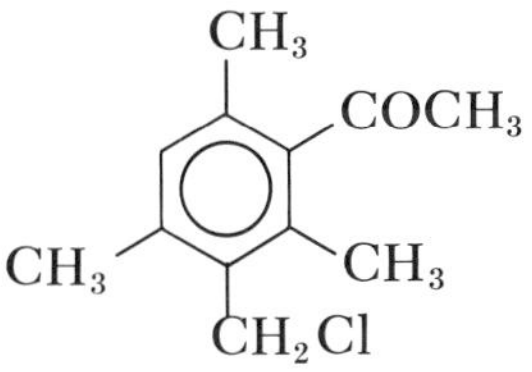 [structure: with CH_3, $COCH_3$, CH_3, CH_3, CH_2Cl]

24. 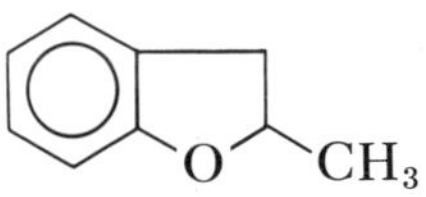 [structure: benzofuran with O, CH_3]

25. 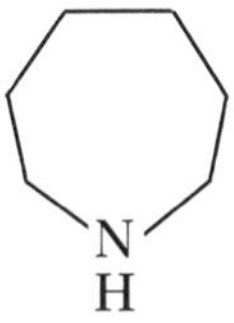 [structure: azepane with N, H]

CHAPTER 12

FREE RADICAL CHLORINATION OF 2,4-DIMETHYLPENTANE

Although saturated hydrocarbons are inert to most acidic and basic reagents, they can be halogenated in the presence of a free radical initiator. The process is a chain reaction as shown in equations 1 through 5 for chlorination with Cl_2.

$$Cl_2 \xrightarrow{h\nu} 2\ Cl\cdot \qquad (1) \qquad \text{initiation}$$

$$\left. \begin{array}{l} Cl\cdot + R{-}H \longrightarrow H{-}Cl + R\cdot \qquad (2) \\ R\cdot\ + Cl_2 \longrightarrow R{-}Cl + Cl\cdot \qquad (3) \end{array} \right\} \quad \text{propagation}$$

$$\left. \begin{array}{l} R\cdot\ + Cl\cdot \longrightarrow R{-}Cl \qquad (4) \\ R\cdot\ + R\cdot \longrightarrow R{-}R \qquad (5) \end{array} \right\} \quad \text{termination}$$

Another method of chlorination involves the use of sulfuryl chloride and benzoyl peroxide ($C_6H_5\overset{\overset{\displaystyle O}{\|}}{C}OO\overset{\overset{\displaystyle O}{\|}}{C}C_6H_5$). At relatively low temperatures, the O—O bond breaks to form two benzoate radicals and the propagation steps are:

$$Cl\cdot + RH \longrightarrow HCl + R\cdot$$

$$R\cdot\ + ClSO_2Cl \longrightarrow RCl + SO_2 + Cl\cdot$$

When a molecule contains more than one type of hydrogen atom, as in isopentane, a mixture of alkyl chlorides can result from monochlorination. The observed composition of the mixture is not the one that would be predicted statistically on the basis of random attack of Cl· on the twelve hydrogens (see table).

$$\begin{array}{c} H_3C \\ {\scriptstyle 1} \end{array} \!\!\diagdown \begin{array}{c} \\ CH{-}CH_2{-}CH_3 \\ {\scriptstyle 2\quad 3\quad 4} \end{array} \xrightarrow[h\nu]{Cl_2} \text{mixture of chloro-2-methylbutanes}$$
$$\begin{array}{c} H_3C \\ {\scriptstyle 1} \end{array} \!\!\diagup$$

ISOPENTANE CHLORINATION PRODUCTS

ISOMER:	1-CHLORO	2-CHLORO	3-CHLORO	4-CHLORO
Observed % of mixture:	34%	22%	28%	16%
Statistical prediction:	50%	8%	17%	25%

Specifically, too little of the primary chlorides are obtained, and too much of the secondary and tertiary. Analysis of the data suggests that the relative rates of abstracting a hydrogen atom from positions 1 through 4 of 2-methylbutane are 1.0:4.0:2.5:1.0, respectively; i.e., $3° > 2° > 1°$. This order, which is the reverse of what would be predicted if the attack were directed by steric effects, is determined by the relative stabilities of the resulting alkyl radicals.

To test the generality of these substitution ratios 2,4-dimethylpentane is to be chlorinated in this experiment, and the products analyzed by VPC.

EXPERIMENTAL PROCEDURE

CAUTION: Sulfuryl chloride vapors are irritating to the eyes and nose, and the liquid causes burns if spilled on the skin.

To reduce hazards *somewhat*, the sulfuryl chloride will be provided in a CCl_4 solution (32 g/100 ml solution). Obtain 1 ml of this solution in a clean dry 18 × 150 mm test tube. To this solution add by pipet 0.5 ml of 2,4-dimethylpentane (d 0.67), and 12.5 mg of benzoyl peroxide (see comparison sample). Attach a drying tube containing KOH pellets and clamp the test tube in a bath of water (400 ml beaker), submerged to a depth of about 1 inch. Around the upper part of the test tube wrap two loops of condenser tubing, hold this with a rubber band or test tube clamp, and connect one end to the water outlet. Pass a gentle stream of water through this loop and into the trough to provide a condensing surface for vapors (Fig. 12.1).

Heat the bath with a burner to 75° and maintain this temperature within ±2° for 30 minutes. After this time, disassemble, cool the solution, add 3 ml of water, stir, and then add 1 ml of sodium bicarbonate solution. Stir the mixture and then transfer the carbon tetrachloride layer with a pipet to a 10 × 75 mm test tube. Add a few grains of Na_2SO_4 to dry. A sample of this solution is then removed with a microliter syringe for gas chromatography. The order of elution of the compounds is (air), 2,4-dimethylpentane, CCl_4, 2-chloro-2,4-

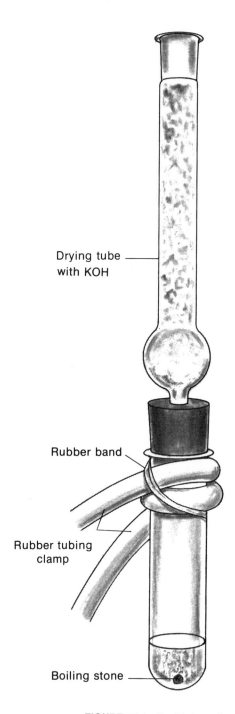

Drying tube
with KOH

Rubber band

Rubber tubing
clamp

Boiling stone

FIGURE 12.1 Test tube reflux set-up.

dimethylpentane, 3-chloro-2,4-dimethylpentane and 1-chloro-2,4-dimethylpentane.

By triangulation or cutting and weighing (Chapt. 5), determine the areas of the dimethylpentane and chlorodimethyl-pentane peaks. (A change in attenuation will be required to keep all the peaks on scale and of measurable size. Consult your instructor for the exact settings.) Assuming that the areas are proportional to the weights of the various compounds, calculate the total per cent of chlorination and the per cent distribution among the chlorinated products.

QUESTIONS

⋆1. Using the relative reactivity of 1°:2°:3° hydrogens observed with 2-methylbutane (1.0:2.5:4.0), predict the composition of the monochlorination product mixture from
- a. propane
- b. butane
- c. isobutane
- d. 2,4-dimethylpentane

⋆2. The products of chlorination of cyclohexane (RH) with sulfuryl chloride initiated by benzoyl peroxide are cyclohexyl chloride, HCl, SO_2, chlorobenzene and CO_2. Give equations for the overall reaction (initiation, propagation, and termination steps) which account for these products.

3. How does your observed isomer distribution compare with that which you calculated in question 1d? Comment on the results. A set of molecular models will be useful here; reference may be made to a paper by M. S. Newman in *J. Am. Chem. Soc.*, **72**, 4783, 1950.

4. In Figures 12.2 to 12.4 are the nmr spectra of the three chlorodimethylpentanes obtained in this experiment. Identify the compounds by their spectra and explain your reasoning.

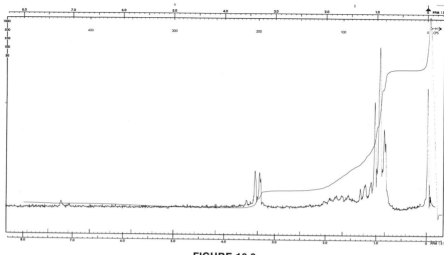

FIGURE 12.2

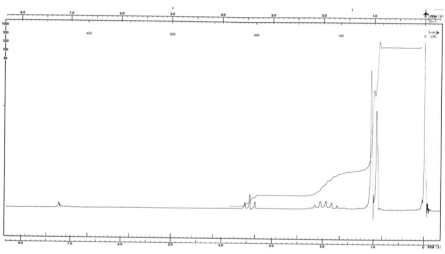

FIGURE 12.3

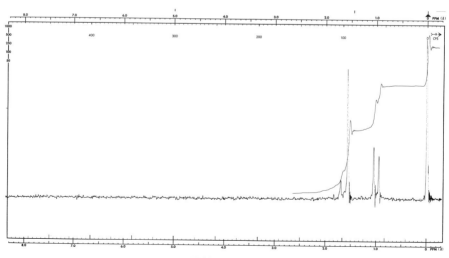

FIGURE 12.4

CHAPTER 13

DEHYDRATION OF
2-METHYLCYCLOHEXANOL

bp 165–168°   bp 110°   bp 104°
d 0.93

Elimination reactions of alcohols and alcohol derivatives are frequently an important step in synthesis, since alcohols are available in wide variety. With secondary and tertiary alcohols, direct acid-catalyzed dehydration is a simple and convenient procedure. A problem is encountered, however, when dehydration can occur in two directions, giving rise to either or both of two products. This situation is encountered with 2-methylcyclohexanol, where 1- or 3-methylcyclohexene may be produced. In the experiment below the actual course of the reaction will be determined by VPC analysis of the product.

Either sulfuric or phosphoric acid may be used to dehydrate alcohols. A disadvantage of the former, particularly in this experiment, is that H_2SO_4 is a strong enough acid to reprotonate the double bond of the product and isomerize it to other methylcyclohexenes.

EXPERIMENTAL PROCEDURE

To a 50 ml round-bottom flask add 10 ml of 2-methylcyclohexanol, 3 ml of 85 per cent phosphoric acid, and a boiling chip. Attach the flask to a fractional distillation assembly (Fig. 4.5). Slowly heat the contents of the flask to boiling and distill out the product. The vapor temperature should be kept

111

below 96° by regulating the rate of heating. Continue distilling until 8 to 10 ml of liquid have been collected.

Transfer the distillate to a separatory funnel and remove the water layer. Transfer the organic phase to a clean, dry Erlenmeyer flask and dry the liquid for 10 minutes over $CaCl_2$.

Inject 0.5 μl of the dried liquid onto a nonpolar column of a gas chromatograph. Before doing so, be sure to fill and empty the syringe several times to clean out the previous contents, and to draw approximately 1 μl of air into the syringe after the liquid. Insert the needle as far as possible into the injection port before injecting the sample; in this way it will be injected directly onto the column.

Mark the point of injection on the recorder chart paper and measure accurately the distance (or time) from this point to the air peak and the olefin peak(s). By comparing the latter with the retention times of 1-methylcyclohexene and 3-methylcyclo-hexene (VPC's of these compounds will be provided), deter-mine the identity of the dehydration product(s). If both are present, determine the approximate composition by triangula-tion. Explain your results.

Calculate the R_F value(s) of the methylcyclohexene(s) and also the number of theoretical plates in the column (see Chapt. 5).

QUESTIONS

*1. Write equations for the mechanism of dehydration of 2-methylcyclohexanol, showing how both possible products may be formed.

*2. Give equations to show how sulfuric acid can convert 4-methylcyclohexene into a mixture of four isomeric C_7H_{12} compounds (including starting material). Predict the relative amounts of each isomer.

*3. What product(s) would you expect from H_3PO_4 treat-ment of (a) 1-methylcyclohexanol, (b) 4-methylcyclohexanol, (c) cyclohexylmethanol?

4. The 2-methylcyclohexanol used in this experiment is actually a mixture of two isomers. Is it possible or likely that this makes any difference in the product composition? Explain your answer.

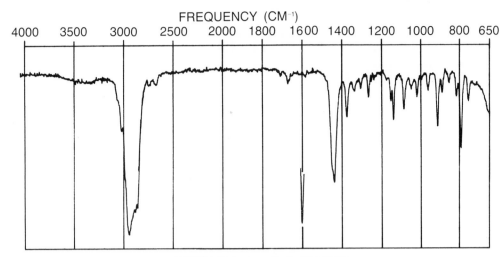

FIGURE 13.1 1-methylcyclohexene.

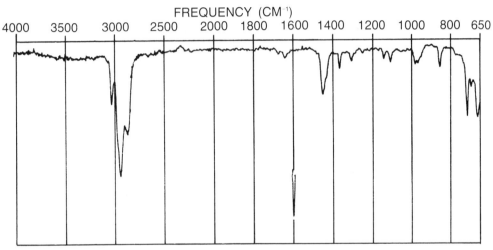

FIGURE 13.2 3-methylcyclohexene.

5. Infrared spectra of the two methylcyclohexene isomers obtained in this experiment are given in Figures 13.1 and 13.2. Assign vibrational modes for as many peaks as possible in both spectra.

As an alternative to VPC analysis, the relative intensities of peaks in the infrared spectrum of a mixture may be used to determine its composition. Which peaks in Figures 13.1 and 13.2 could be used to analyze a mixture of these isomeric methylcyclohexenes?

Reference

R. L. Taber, and W. C. Champion, *J. Chem. Educ.*, **44**, 620 (1967).

14

DIELS-ALDER REACTION: SEPARATION OF *CIS* AND *TRANS* PIPERYLENE

The Diels-Alder reaction is an important synthetic tool for building cyclic systems. The reaction comprises the cycloaddition of a conjugated diene and another unsaturated compound, the dienophile. The dienophile usually contains a double or triple bond conjugated with an electron-withdrawing group or groups. A typical synthetic application of the Diels-Alder reaction is the condensation of butadiene and 4-methoxy-2,5-toluquinone; this is the first step in a classical synthesis of the steroid hormones.

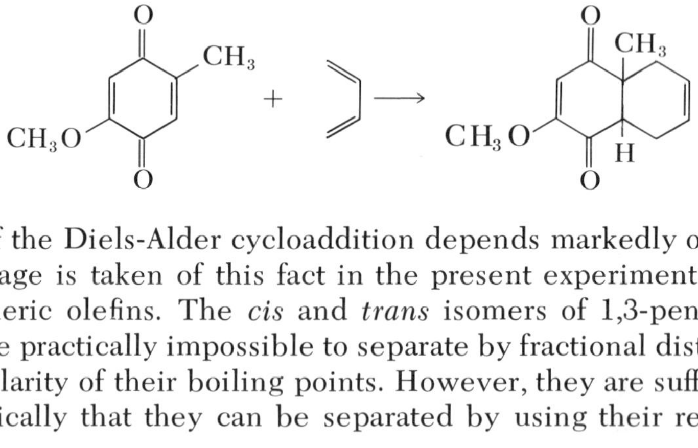

$$(14.1)$$

The rate of the Diels-Alder cycloaddition depends markedly on steric effects. Advantage is taken of this fact in the present experiment to separate two isomeric olefins. The *cis* and *trans* isomers of 1,3-pentadiene (piperylene) are practically impossible to separate by fractional distillation due to the similarity of their boiling points. However, they are sufficiently different chemically that they can be separated by using their reactivity

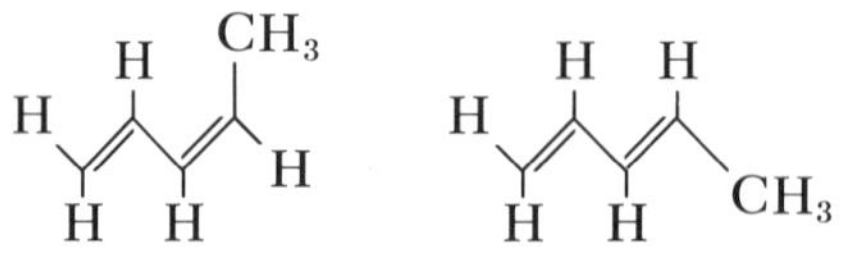

cis
bp 43.5°

trans
bp 42°

114 with various reagents. The best method for obtaining one of the isomers from a mixture of the two is by removing the other isomer as its Diels-Alder adduct with maleic anhydride. The isomer which does not react can be isolated by simple distillation from the reaction mixture.

$$(14.2)$$

mp 61°

EXPERIMENTAL PROCEDURE

In a 100 ml round-bottom flask place 27 g of maleic an-hydride and 0.5 g of picric acid. Attach a reflux condenser and warm the mixture on a steam bath until the maleic anhydride melts, then turn off the steam. From a dropping funnel sus-pended above the condenser (do not seal the system!), add dropwise 40 ml (d. 0.68) of commercial piperylene at a rate which maintains a gentle reflux. Swirl the flask during addition to ensure mixing. After the addition is complete, heat the flask under the condenser in a water bath heated to 50 to 60° by the steam bath (no flames!).

While waiting for the reaction to be completed, check the composition of a sample of the commercial piperylene by VPC. On a nonpolar (silicone) column at room temperature, the isomers elute in the order of their boiling points. Record their retention times or R_F values relative to air, and estimate by triangulation the per cent composition of the mixture.

After a total of one hour of refluxing, reassemble the apparatus for a simple distillation and distill out the unreacted piperylene. Compare the VPC of the distillate with that of the starting mixture, and identify the isomers as to reactivity with maleic anhydride. If some question exists as to the identity (due to the similarity of R_F values), positive identification can be made by mixing a few drops of the mixture and the distillate, and observing which peak (*cis* or *trans*) in the mixture is en-hanced on VPC.

Transfer the residue from the distillation flask to a 250 ml separatory funnel containing 75 ml of benzene and 50 ml of water. Shake well and discard the aqueous layer; if an emulsion forms add 10 ml of ether to hasten separation. If any solid ma-terial is present at the interface, this should be removed by filtering that part of the solution in which the solid is suspended. Rinse the organic layer with 50 ml of water and with three 25-ml portions of 10% $NaHCO_3$ solution. Dry over $MgSO_4$ or Na_2SO_4, filter, and concentrate to approximately 60 ml on a steam bath. Let the solution cool to room temperature and then add 100 ml of hexane in 20-ml portions, swirling the solution after each addition; continue until the solution becomes turbid and then initiate crystallization by scratching or seeding. After crystallization has begun, add the remaining hexane. Collect the product and recrystallize if necessary from hexane-benzene (2:1). Report the melting point and yield, and submit the compound to your instructor.

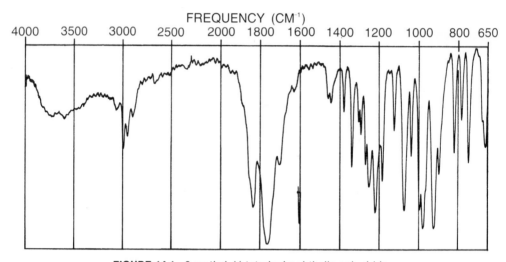

FIGURE 14.1 3-methyl-Δ^4-tetrahydrophthalic anhydride.

QUESTIONS

1. Which isomer of piperylene did not form an adduct with maleic anhydride? Speculate as to why.

2. Given the stereochemistry of the isolated adduct shown in Equation (14.2), draw the configuration of the reactants in the transition state of this Diels-Alder reaction.

3. The other isomer of piperylene may be obtained in a pure state by the following series of reactions. The mixture of *cis* and *trans* isomers is heated in a sealed tube with sulfur dioxide. After the tube has been cooled and opened, a compound $C_5H_8SO_2$ is isolated. Heating this compound causes it

to decompose to SO_2 and piperylene. The piperylene so obtained is the opposite isomer to that isolated in this experiment. Postulate a structure for the $C_5H_8SO_2$ compound.

Reference

R. L. Frank, R. D. Emmick, and R. S. Johnson, *J. Amer. Chem. Soc.,* **69,** 2313 (1947).

15

FRIEDEL-CRAFTS REACTION:
p-DI-t-BUTYLBENZENE

Friedel-Crafts alkylation is the most general method for introducing alkyl groups into a benzene ring. There are definite practical limitations, however. Since the reaction involves the development of positive charge on the alkyl carbon atom, rearrangement of the alkyl group may occur. Moreover, the ring becomes more susceptible to substitution (more nucleophilic) as alkyl groups are introduced, and there is thus a tendency for polysubstitution. Finally, "scrambling" of alkyl groups can take place when a thermodynamically more stable product can arise from one that is formed initially in a faster reaction.

EXPERIMENTAL PROCEDURE

In this experiment, benzene is alkylated by t-butyl chloride. One of the products will be isolated; others can be observed by VPC.

Procedure. In a thoroughly dry 125 ml Erlenmeyer flask place 20 ml of t-butyl chloride and 10 ml of benzene. Arrange a trap for HCl gas by connecting a length of rubber tubing from a one-hole rubber stopper in the flask to a glass funnel inverted just below the surface of a beaker containing water. This is a simple and safe way to handle acid fumes; the system is sealed

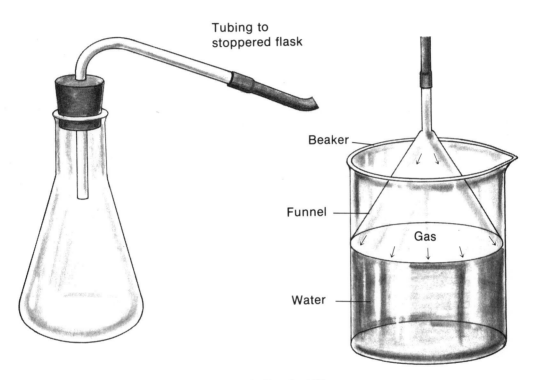

FIGURE 15.1 Trap for HCl gas.

with water to prevent escape of HCl but the water cannot back up into the reaction flask because of the large empty volume of the funnel.

Obtain 1 g of anhydrous aluminum chloride in a stoppered 10 × 75 mm test tube. Minimize exposure of this compound to air; the proper amount can be estimated by comparison with a weighed sample prepared by the instructor. Cool the benzene-*t*-butyl chloride mixture in an ice bath (clamp the flask loosely to prevent tipping). Add about one-fourth to one-third of the AlCl₃ and swirl the flask in the bath. After bubbling has occurred for 4 to 5 minutes, add the rest of the AlCl₃ in two separate portions during the next 10 to 15 minutes. (Restopper the flask as quickly as possible, and rinse your fingers, since brief exposure to HCl fumes cannot be completely avoided.)*

When the reaction begins to subside remove the ice bath and allow the mixture to warm to room temperature. Unstopper and add ice-cold water (from the bath). Then add about 20 ml of ether and transfer the contents of the flask to a separatory funnel; rinse the flask with ether. Separate the ether layer, wash with water, dry with MgSO₄, and evaporate the solution in an Erlenmeyer flask. Crystallize the *p*-di-*t*-butylbenzene from about 20 ml of methanol, collect and dry the product, and record its weight and melting point.

FIGURE 15.2

---

*In a larger scale reaction, it would be quite undesirable to open the system repeatedly to add the AlCl₃. A very satisfactory procedure for adding a solid reagent in portions to a moisture-sensitive reaction is shown in Figure 15.2. A wide test tube containing the solid is connected to a neck of the flask by a short length of Gooch crucible tubing. The solid can be sealed from the reaction by collapsing the tubing, and then added simply by tipping up the test tube and shaking in the reagent.

QUESTIONS

*1. What monoalkylated product(s) would you expect to obtain if *n*-butyl chloride were used instead of *t*-butyl chloride?

*2. Under forcing conditions, it is possible to place *seven* methyl groups on a benzene ring by alkylation. The product is a stabilized carbonium ion salt; suggest the structure of this product.

*3. Calculate the molar ratio of *t*-butyl chloride and benzene used in this experiment.

4. VPC analysis of the crude reaction mixture from this experiment shows, in addition to unreacted benzene and *p*-di-*t*-butylbenzene, three other products (see Fig. 15.3). Suggest structures for these by-products, and write equations for their formation.

5. How would you change the experimental conditions in order to obtain (a) compound 1, and (b) compound 3, as the major product of the reaction?

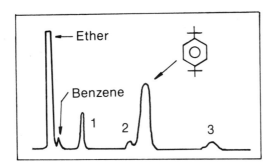

FIGURE 15.3 VPC of Friedel-Crafts reaction mixture.

16

PREPARATION OF CYCLOHEXYL CHLORIDE

The conversion of alcohols to alkyl chlorides can be accomplished by several procedures. With primary and secondary alcohols, thionyl chloride or phosphorus halides are often used; these halides can also be obtained by heating the alcohol with concentrated hydrochloric acid and anhydrous zinc chloride. Tertiary alcohols are converted to the chloride by concentrated hydrochloric acid alone, in some cases without heating. Cyclohexanol is somewhat exceptional in its reactivity. Although it is a secondary alcohol, it is easily converted to the chloride by refluxing in concentrated hydrochloric acid.

These reactivity differences are the basis of a qualitative test for 1°, 2°, and 3° alcohols (Lucas test). The reagent is a solution of zinc chloride in hydrochloric acid (100 g $ZnCl_2$ + 75 g concentrated HCl). The alcohol is added to this solution. The test depends on the separation of the alkyl chloride from solution to give an oily layer or turbid suspension, and is restricted to alcohols that are completely miscible in the reagent, generally those containing no more than about 6 carbon atoms. Tertiary or allylic alcohols react in a few seconds, and secondary alcohols in 10 to 15 minutes. Primary alcohols give no reaction.

An important concern in the preparation of alkyl chlorides and bromides from the alcohols is contamination of the product with unreacted alcohol. This is particularly serious if the halide is to be used in a Grignard reaction. Careful fractional distillation usually effects removal of the alcohol from a chloride, but often not in the case of an alkyl bromide. To completely remove all alcohol, the chloride or bromide is washed with concentrated sulfuric acid, followed by water. The alcohol is converted to a water-soluble alkyl hydrogen sulfate which is extracted by water.

EXPERIMENTS

A. Preparation of Cyclohexyl Chloride

Apparatus. Since HCl fumes are evolved in this reaction, provision must be made to trap gases from the condenser. This can be done with the inverted funnel trap, as used in the Friedel-Crafts reaction, attached to the tube adapter at the top of the condenser. A more efficient arrangement for a prolonged reflux period is to run the rubber tubing from the tube adapter to a scrubber as shown in Figure 16.1, in which the exit gases are discharged next to the outlet condenser water. This scrubber must be used if you find that the funnel trap is not handling the fumes adequately.

The scrubber consists of a bent adapter with a two-hole cork or rubber stopper with two short lengths of glass tubing. The adapter is arranged to discharge into the sink or trough.

CAUTION: The adapter may tend to fill with water, especially if too much water is going through the condenser.

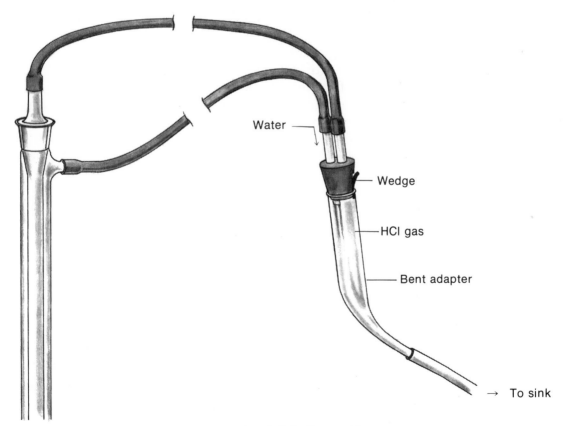

Water

Wedge

HCl gas

Bent adapter

→ To sink

FIGURE 16.1 Gas scrubber.

If water backs up to the tube venting the HCl, it will suck back into the reaction. This problem can be overcome by providing a small vent to the atmosphere between the stopper and adapter, using a paper match as a wedge.

Procedure. In a 250 ml round-bottom flask place 30 ml of cyclohexanol and 120 ml of concentrated hydrochloric acid. A reflux condenser with tubing to the gas trap or scrubber is attached. The flask is then heated (boiling stone!) with a burner (wire gauze and ring) to gentle boiling. Care should be taken during the initial heating not to reflux too vigorously, or product may be lost by entrainment with HCl. After 20 minutes, increase the heating rate to maintain vigorous boiling and turbulence (check that the joint at the top of the condenser does not become warm due to too rapid reflux). Maintain refluxing for a total of $2\frac{1}{4}$ to $2\frac{1}{2}$ hours(see question 6).

(During this time, carry out the experiment in part B or other experiments in the schedule.)

After cooling, pour the reaction mixture into a separatory funnel and remove the lower aqueous layer as completely as possible. Then add 10 ml of concentrated sulfuric acid to the chloride and shake until the layers are thoroughly mixed (see question 7).

CAUTION: Concentrated H_2SO_4 can cause severe skin burns. Be sure that the stopper and stopcock are securely seated, wipe up any leakage of the lower layer, and wash your hands after the extraction.

Allow 10 to 15 minutes for the layers to separate, and slowly drain off the sulfuric acid (several minutes are required for the viscous layer to completely flow down the walls). The chloride is then washed with 20 ml of water. An emulsion will form at this point which requires some time for separation. This is a convenient stopping point; store in a stoppered flask.

Separate the aqueous phase and wash the turbid chloride layer, which still contains some emulsified water, with 10% Na_2CO_3 solution. The layers should separate more rapidly, but some minutes may be required. A little salt can be added to hasten separation. Wash again with water, dry the product by swirling with calcium chloride and filter through a small cotton plug into a 50 ml round-bottom (distilling) flask. Distill, without a fractionating column, collecting material with boiling point 138 to 142° in a tared receiver. If any traces of water are noted in the distillate (insoluble droplets or turbidity) the chloride must be further dried and redistilled, since absolutely dry material is required for the next reaction. Calculate the number of moles and yield, and the weight of magnesium required for conversion to the Grignard compound.

B. Lucas Test

Obtain 10 ml of the HCl-ZnCl$_2$ reagent and distribute it equally in five 10×75 mm test tubes. To each tube add about 0.2 ml of one of the following alcohols: (a) 1-butanol, (b) 2-butanol, (c) *t*-butyl alcohol, (d) cyclohexanol, and (e) an unknown C—5 alcohol. Record the results and make a conclusion as to whether the unknown alcohol is primary, secondary, or tertiary.

QUESTIONS

⋆1. Consult a source such as *Synthetic Organic Chemistry* by Wagner and Zook or *Organic Syntheses* and find the procedure and conditions used to prepare the following halides from the alcohols. Note any comments on the advantages and disadvantages of alternative methods.
   a. *n*-Hexyl chloride
   b. 1-Chloro-2-ethylbutane
   c. 1-Chloro-2-methylcyclopentane
   d. 2-Methyl-2-chloropentane.

⋆2. Suggest the method and approximate conditions that you feel would be appropriate to prepare 1-ethyl-1-chlorocyclohexane.

⋆3. In the Lucas reagent, zinc chloride functions as a Lewis acid. Explain what this means, and specifically how this type of catalyst exerts its effect.

⋆4. Why can't distillation be used to separate unreacted alcohol from an alkyl bromide?

5. Does the reactivity of cyclohexanol, as discussed in the introduction, parallel the reactivity observed in the Lucas test?

6. Why does a phase separation occur during the reaction? What are the principal organic components in each phase?

7. Write a balanced equation for the reaction that occurs when sulfuric acid is added to the crude cyclohexyl chloride. The product of this reaction is primarily responsible for the emulsion in the aqueous wash step. Explain why emulsion formation occurs.

8. If you were carrying out a large scale preparation of cyclohexyl chloride from cyclohexanol, how might the procedure be modified to increase the yield? What specific result would the changes be expected to have?

CHAPTER 17

REACTIVITY OF HALIDES

Alkyl halides or alkyl sulfonate esters serve as the reactant or "substrate" in almost all nucleophilic substitution reactions. These include a number of useful preparative reactions leading to a wide variety of products. The mechanisms of nucleophilic substitution have been very extensively studied and progressively refined to account for the effects of solvent, added electrolytes, structure of the halide, nature of the leaving group, and other factors.

Nucleophilic substitution reactions can be viewed in a highly oversimplified way as occurring with varying contributions of two limiting mechanisms:

$$S_N2 \quad Nuc:^- \cap \underset{|}{C}-X \longrightarrow Nuc \overset{\delta-}{\cdots} \underset{|}{C} \overset{\delta-}{\cdots} X \longrightarrow Nuc-\underset{|}{C} + X^-$$

$$S_N1 \quad \underset{|}{C}-X \longrightarrow \underset{|}{C}^+ \quad X:^- \xrightarrow{Nuc:^-} Nuc-\underset{|}{C} \quad and \quad \underset{|}{C}-Nuc$$

The optimum conditions for these two extremes are approximated by two reagents which can be used to examine the mechanism of nucleophilic substitution in alkyl halides in a qualitative way. A solution of sodium iodide in acetone represents essentially the limiting conditions for S_N2 displacement; ethanolic silver nitrate permits the observation of S_N1 solvolysis.

EXPERIMENTS

In these experiments a series of alkyl halides will be tested for the rate of substitution by the S_N1 and S_N2 processes. In both cases, the rate of formation of a visible precipitate provides an indication of the relative importance of the two mechanisms. With NaI in acetone, the precipitate is

NaBr or NaCl, both of which have very low solubility in this solvent. With alcoholic silver nitrate, the silver halide precipitates.

> **Procedure.** Label two series of 6 clean dry test tubes (10 × 75 mm can be used for one series and 16 × 150 for the other) from 1 to 6. In each series place 0.2 ml of the following halides: (1) *n*-butyl chloride, (2) *n*-butyl bromide, (3) *sec*-butyl chloride, (4) *t*-butyl chloride, (5) α-chloroacetone (ClCH$_2$COCH$_3$), and (6) crotyl chloride (CH$_3$CH=CHCH$_2$Cl).
>
> Obtain about 15 ml each of solutions of 15% NaI in acetone and 1% ethanolic AgNO$_3$ from the side shelf. In one series add to each test tube in turn, noting the time, 2 ml of the NaI-acetone solution. Observe any rapid reactions, then add to the other series in turn 2 ml of the AgNO$_3$ solution, noting time of addition. In the latter series, it is well to note the time both for the first turbidity and for a definite precipitate.
>
> After about 5 minutes, place any tubes in the NaI-acetone series that do not contain a precipitate in a 50° water bath (note time).
>
> Summarize your findings and conclusions. Account for the differences in reactivity observed and draw correlations of the relative rate with the structure of the alkyl radical and the nature of the leaving group.

QUESTIONS

⋆1. Write equations for the preparation of the following compounds by nucleophilic substitution reactions:
   a. 1-pentyne
   b. methyl phenyl sulfide
   c. ethyltrimethylammonium bromide
   d. 2,4-dimethyl-2-pentanol
   e. isobutyronitrile

⋆2. Certain nucleophiles are termed "ambident" because they possess more than one site for coordination of an electrophilic carbon atom. A few examples are nitrite ion, cyanide ion,

$$\text{enolate ions } (R-C\overset{\displaystyle O}{\overset{\|}{=\!=\!=}}C-R')^- \text{ and sulfinate anions } (R-SO_2-)^-.$$

Write the structures of the two products that could be obtained in the reaction of each of these nucleophiles with ethyl iodide.

3. For each of the following compounds indicate whether nucleophilic substitution by an S_N1 or an S_N2 process would be faster. For compounds that would be faster by the S_N2 process, indicate whether they would be faster or slower than *n*-butyl chloride. For reactions that would be faster by an S_N1 process, indicate whether they would be faster or slower than *t*-butyl chloride. Give reasons for your predictions.

    a. 3-chloro-3-ethyl-2-methylpentane

    b. 3-chloro-2-butanone

    c. 1-chlorobicyclo[2.2.2]octane

    d. 2-chloro-2-phenylethane

    e. ethyl iodide

4. Suggest an experimental procedure by which the relative S_N2 reactivities of 1-, 2-, 3-, and 4-iodooctanes could be determined.

CHAPTER 18

GRIGNARD REACTION: CYCLOHEXYL PHENYL CARBINOL

Grignard reagents play a commanding role in organic synthesis. These compounds can be adapted to the preparation of a large variety of functional systems, and the formation and reactions of organomagnesium derivatives comprise one of the major uses of alkyl halides in organic synthesis.

Alkyl bromides are most frequently employed for the preparation of Grignard reagents, since these are usually as accessible as the corresponding chloride and are somewhat more reactive. Cyclohexyl chloride is easily converted to the Grignard reagent, however, and is just as suitable as the bromide.

Reaction of the halide and magnesium occurs on the metallic surface and is formally an oxidation of the metal. The reaction is usually carried out in ether solution. The ether solvent functions as a Lewis base, solvating the Grignard reagent and permitting it to diffuse away from the metal.

The major pitfall in preparing a Grignard compound is the presence of alcohol or any other active hydrogen compound:

$$HY + RMgX \rightarrow RH + MgXY$$

This reaction consumes and wastes the organomagnesium halide, and it **129** can also seriously complicate its formation, since the product MgXY can coat the surface of the metal and prevent attack of the halide.

Reaction of the Grignard reagent with carbonyl compounds is usually carried out in the same flask in which the organometallic solution is prepared. After addition of the aldehyde or ketone, the alkoxymagnesium halide is hydrolyzed to liberate the alcohol. Ammonium chloride is used to complex the magnesium ion, thus avoiding an excess of strong acid.

EXPERIMENT

Cyclohexyl chloride prepared in the previous experiment is to be converted to the Grignard reagent and thence to cyclohexyl phenyl carbinol.

Procedure. A 500 ml 3-neck flask is set up with a condenser in a side neck, dropping funnel in the center neck, and a stopper in the third neck. Clamp the flask securely by the center neck, but do not use a ring; allow room to slide an ice bath under the flask. *All glassware must be completely dry.* Fit a drying tube (see Fig. 12.1) containing calcium chloride in the top of the condenser with a rubber sleeve and tube adapter.

Place in the flask an amount of magnesium turnings equivalent in moles to the cyclohexyl chloride available, and add 20 ml of anhydrous ether. Fill the condenser jacket with water and turn off the flow. The Grignard reaction is initiated by adding 2 to 3 ml of the chloride, with a pipette, in a concentrated pool at the surface of the metal (lower the tip of the pipette below the surface of the ether and then release the halide). With a stirring rod, gently rub and crush a piece of the magnesium at the point where the chloride was added, and stopper the flask. Warm the flask with warm water or the palm of the hand but do not shake or agitate (a high concentration of the halide at the surface of the magnesium is desired to initiate the reaction). If no visible sign of reaction (bubbling, turbidity) is noted after 5 minutes, add a small crystal of iodine and another 1 ml of the halide. If there is no evidence of reaction after an additional 5 minutes, add a small sample of reacting solution from another student, or one prepared separately in a dry test tube from a few chips of magnesium, 1 ml of ether, and 1 ml of cyclohexyl chloride.

When the reaction is actively in progress (heat evolved, turbulence), begin a flow of condenser water. Dilute the remaining chloride with anhydrous ether (3 ml per gram) and place this solution in the dropping funnel. Add the chloride solution dropwise, with occasional swirling of the flask, at a rate to maintain steady refluxing. After all the halide has been added, warm the solution on the steam bath as needed to maintain reflux for another 20 minutes. At this point the solution should be a black color (due to finely dispersed impurities from the magnesium) with a few bits of metal remaining.

While the Grignard solution is being refluxed, weigh out benzaldehyde equivalent in moles to the amount of cyclohexyl chloride used, mix the aldehyde with three volumes of anhydrous ether, and place in the dropping funnel. When the Grignard formation is complete, cool to room temperature, and then add the aldehyde solution dropwise at a rate to maintain reflux, swirling to disperse the aldehyde as it is added. Intermittent cooling with an ice bath will permit more rapid addition. The rate can be increased as the reaction proceeds. Swirl for a few minutes after the last of the aldehyde has been added and pour the reaction mixture into a 250 ml Erlenmeyer flask, rinsing with ether. Avoid transferring chips of unreacted metal, which can interfere in subsequent steps. The flask can be corked and stored at this point.

For hydrolysis, add ether (ordinary—not anhydrous), to replace any lost by evaporation, and a few pieces of ice. Then add saturated aqueous ammonium chloride solution, shaking and cooling as necessary after each 4 or 5-ml portion, until the bulk of the gelatinous solid has disappeared. Pour the mixture into a separatory funnel, add a little dilute hydrochloric acid $(1-2N)$, and shake to dissolve any remaining solid. Wash the ether layer with water and then with two 10-ml portions of 30 per cent (saturated) sodium bisulfite solution (shake for at least 1 minute with each portion of $NaHSO_3$), and finally wash again with water (see question 5). Dry $(MgSO_4)$ and evaporate the ether solution in a tared 25×100 mm test tube to an oily residue. Cool in an ice bath, scratch, and allow the oil to crystallize. If crystallization does not occur readily, add 1 ml of pentane. Set up a Hirsch funnel for suction filtration with the aspirator turned off. When crystallization is judged to be complete, transfer the slush to the funnel, and then apply suction. (If suction is applied initially, evaporation of solvent from the filtrate will cause the funnel to become clogged.)

Recrystallize the crude alcohol by dissolving it in an equal volume of pentane (or low-boiling petroleum ether); record the melting point and weight of the pure material and submit it to your instructor.

QUESTIONS

⋆1. Write equations for the reaction of cyclohexylmagnesium chloride with the following compounds: acetone, methyl formate, ethanol, carbon dioxide, ethylene oxide.

⋆2. Give a detailed procedure, with amounts of reagents, and so forth, for the conversion of 15 g of 1-bromoheptane to 3-decanol using a Grignard reaction.

⋆3. As with other Grignard reagents derived from secondary halides, cyclohexylmagnesium chloride undergoes a side reaction in certain carbonyl addition reactions. With cyclopentanone, for example, cyclopentanol and cyclohexene are formed together with 1-cyclohexylcyclopentanol. Write a balanced equation for the reaction leading to cyclopentanol and suggest a mechanism by which this reaction could occur.

4. In a reaction of cyclohexylmagnesium chloride with 2-butanone, the Grignard reagent was prepared from 20 g of cyclohexyl chloride and was then treated with a solution of 10 g butanone in 50 ml of ether which contained 1 per cent by weight of water. Calculate the amount of cyclohexane (in g) that would be produced and the maximum theoretical yield of secondary alcohol corrected for loss due to cyclohexane formation.

5. What is the reason for washing the crude product with $NaHSO_3$ solution? Write an equation for the reaction that occurs.

CHAPTER 19

PREPARATION AND STEREOCHEMISTRY OF BICYCLIC ALCOHOLS

Reactions and interconversions of compounds in the monoterpene series have been of great importance in studying the mechanism of carbonium ion rearrangements and the stereoselectivity of various reagents and synthetic reactions. Typical monoterpenes are the ketone camphor and the corresponding epimeric alcohols, borneol and isoborneol. A related series of compounds, not occurring in nature, has also been extensively studied; these are the norbornane derivatives **1–4.** *

The norborneols **2** and **3** can be prepared by two of the major synthetic reactions for alcohols: (a) hydration of alkenes and (b) reduction of carbonyl groups. Since separation of the epimeric norborneols is quite difficult, it is important that preparative methods for each isomer proceed with high stereoselectivity. The two reactions in this experiment fill this requirement and are complementary, each giving a different isomer as the major product. Since only two isomers are possible, the outcome of either preparation will establish the steric course of both reactions. The isomer obtained in the hydration can be oxidized to norcamphor (Chapt. 20), supplying the starting material for the other alcohol.

---

*The prefix "nor" in the names of these compounds indicates the presence of three hydrogen atoms in place of the methyl groups in the natural terpenes (the usual usage for this prefix is to denote absence of a single CH$_2$ group from the parent compound). Systematic names for these bridged bicyclic compounds (and the terpenes as well) are based on the parent hydrocarbon bicyclo[2.2.1] heptane; the numbers in brackets designate the number of atoms in each of the three "bridges" of the bicyclic skeleton.

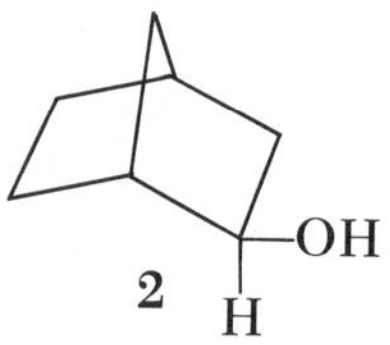

exo-Norborneol, mp 124-126°
Phenylurethane, mp 147°

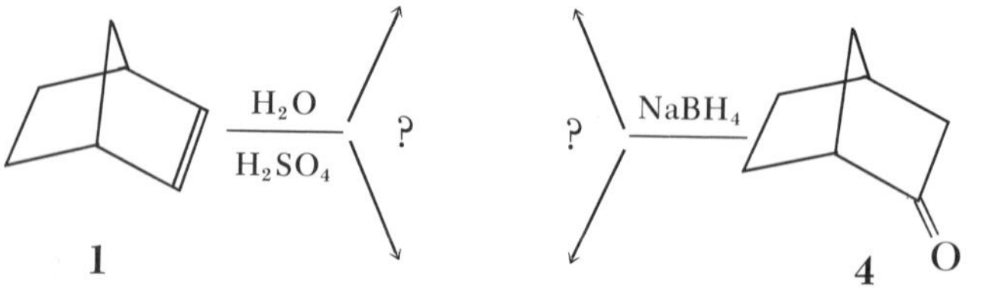

1
Norbornylene

4   O
Norcamphor

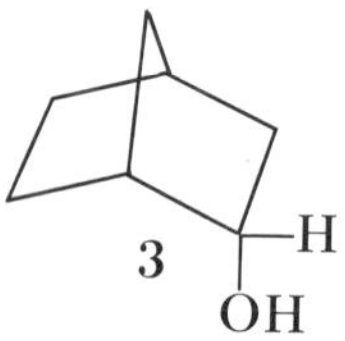

endo-Norborneol, mp 149-150°
Phenylurethane, mp 158°

Hydration of alkenes in the presence of acid proceeds by way of carbonium ions, and mixtures of products are often encountered (see Chapt. 13, question 2). Although carbonium ion rearrangements occur frequently in reactions of terpenes, the hydration of norbornylene (1) proceeds cleanly, and in one steric direction, which is to be determined in this experiment. The mechanism of the reaction and the structure of carbonium ions in bridged bicyclic systems is a subject of some controversy.

The second norborneol is obtained by the hydride reduction of norcamphor. The complex hydrides $NaBH_4$ and $LiAlH_4$ are the most useful reagents available for the conversion of carbonyl compounds to alcohols. Sodium borohydride is the less reactive of the two; e.g., esters and acids are not affected. It is very convenient to use since reactions can be run in aqueous or alcoholic solutions. Reduction of bicyclic ketones such as camphor and norcamphor with these hydrides is quite stereoselective, one of the two diastereomeric alcohols being formed in over nine times the amount of the other. In these rigid cyclic compounds, the stereochemistry of the reduction is controlled by shielding one side of the carbonyl group from attack by the reagent. Thus in camphor, the methyl groups on the one-carbon bridge screen the approach of hydride from the "top" or *exo* side of the two-carbon bridge, and the hydrogen atom is added to the *endo* side, giving the *exo*–alcohol isoborneol.

Camphor   Isoborneol

Since separation of norborneol mixtures is impractical, accurate product ratios can be established only by refined analytical methods. When one isomer greatly predominates, however, as in these experiments, it can easily be isolated in sufficiently pure form to permit identification on the basis of the melting point of the alcohol and, if necessary, a derivative.

EXPERIMENTAL PROCEDURES

Acid-Catalyzed Hydration of Norbornylene

Mix 4 ml of concentrated H_2SO_4 and 2 ml of water, cool in an ice bath to 15 to 20° and add 2.0 g of norbornylene. (NOTE: This hydrocarbon is highly volatile, and it must be added to the acid as soon as it is weighed to avoid loss). Swirl the mixture until all of the hydrocarbon has dissolved, cooling briefly in cold water if the mixture becomes perceptibly warm—do not cool below room temperature. Allow the solution to stand and prepare a solution of 3 g of KOH in 15 ml of water. Cool both solutions in an ice bath and slowly add the KOH solution to the acid reaction mixture. Transfer to a separatory funnel and extract with ether (40-ml and 10-ml portions). Three liquid phases may be present in the separatory funnel at this point; if this occurs, addition of a few ml of water will cause the lower two layers to coalesce. Wash the ether solution with 5 ml of water and 10 ml of $NaHCO_3$ solution, dry ($MgSO_4$), filter into a 125 or 250 ml filter flask and concentrate to an oil on the steam bath (Fig. 3.6). Cool the flask to room temperature, stopper, and evacuate (aspirator) to remove residual solvent.

Sublimation

Sublimation is very effective for purification of the nor-

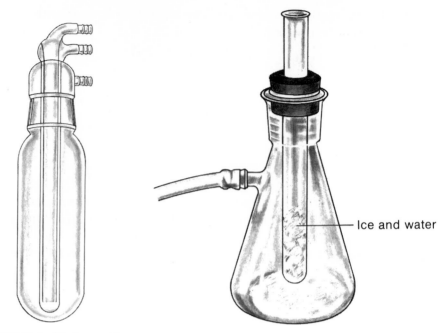

FIGURE 19.1 All-glass sublimer.

FIGURE 19.2 Filter flask–test tube sublimer.

Ice and water

borneol. In this technique, a solid is placed in a vessel which can be evacuated and heated, and in which a cold finger is positioned a short distance above the material to be sublimed. A typical all-glass sublimer is shown in Figure 19.1. For this experiment a very simple sublimer can be set up using locker equipment, as shown in Figure 19.2. A filter flask serves as the subliming vessel and a 18×150 mm test tube as the cold finger. The test tube is fitted with a rubber stopper containing a 16 mm hole. Lubricate the rubber sleeve with glycerine, slip it over the test tube, and adjust the position so that the cold finger is about 2 cm above the bottom of the flask when it is securely seated. Wash off the excess glycerine with water, wipe the test tube clean, and dry thoroughly.

Insert the cold finger, wedge it securely in the neck, and again evacuate the flask with the aspirator. Half fill the cold finger with chipped ice and gently heat the flask on the steam bath until the alcohol has completely sublimed. Cool the flask, disconnect from the aspirator, and pour the water out of the cold finger. Loosen the cold finger from the neck and pull it out gently, holding in a horizontal or slightly inverted position. The crystalline deposit is soft, and if need be, it can be compressed somewhat without crumbling. Scrape the crystals from the test tube, weigh, and determine the melting point in a sealed capillary. (Fill the capillary in the usual way and then heat about 3/4 of an inch below the open end until soft. Pull out and fuse in the flame at the constriction.)

Report which isomer was obtained. To confirm the identification, if necessary, a sample of the alcohol can be converted to the phenylurethane (see p. 227).

Sodium Borohydride Reduction

In a 50 ml Erlenmeyer flask dissolve 1.1 g of norcamphor in 5 ml of ethanol and cool the solution in ice. In a small test tube, dissolve 0.15 g of $NaBH_4$ in 2 ml of water. Add the borohydride solution to the camphor and allow the solution to warm and stand for 10 minutes. Add 6 ml of water and heat the solution for 5 minutes on the steam bath. Cool, extract with 20 ml of pentane, wash with water, dry over $MgSO_4$, and evaporate. When the volume has been reduced to a few milliliters, transfer the solution with a pipet to a 25×100 mm test tube for final concentration and crystallization in order to minimize losses in transferring the solid. At a volume of 1 to 2 ml, chill the solution in an ice bath and transfer the resulting crystalline slush to a Hirsch funnel before connecting to the aspirator, then pull the mother liquor through. Weigh the product, determine the melting point (sealed capillary), and report which isomer was obtained.

QUESTIONS

⋆1. The major source of norbornane derivatives is the Diels-Alder reaction (Chapt. 14). Formulate the preparation of norbornylene, and *endo*-norborneol, using ethylene and vinyl acetate ($CH_2{=}CHO\overset{\displaystyle O}{\overset{\|}{C}}CH_3$), respectively, with a suitable diene.

2. Isoborneol can be prepared (as the acetate) from α-pinene, the major constituent of turpentine. Formulate

mechanisms for the steps in this synthesis as outlined:

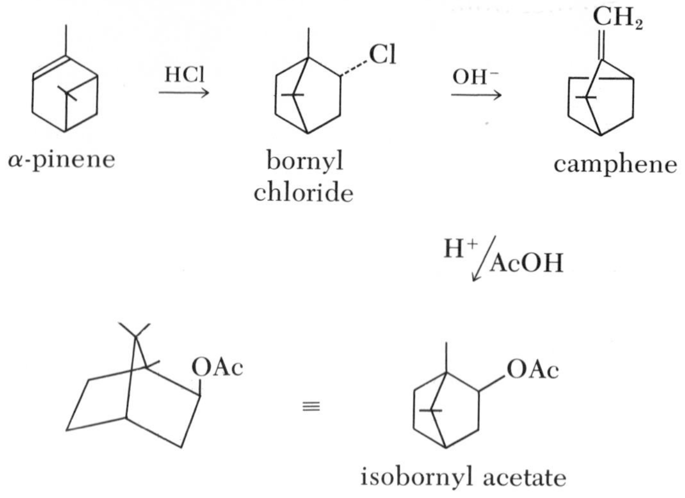

α-pinene   bornyl chloride   camphene

isobornyl acetate

3. Compare the steric course of hydride reductions of camphor and norcamphor. Suggest a reason for the stereochemistry of the norcamphor reaction.

4. According to one viewpoint, the hydration of norbornylene is formulated with the "nonclassical" carbonium ion (i):

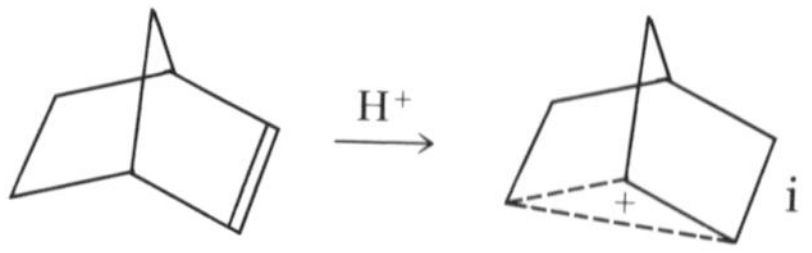

Explain why solvation of this ion would give the observed alcohol isomer.

5. Another method of hydration of olefins is hydroboration-oxidation, which brings about *cis*-addition of water *via* an organoborane:

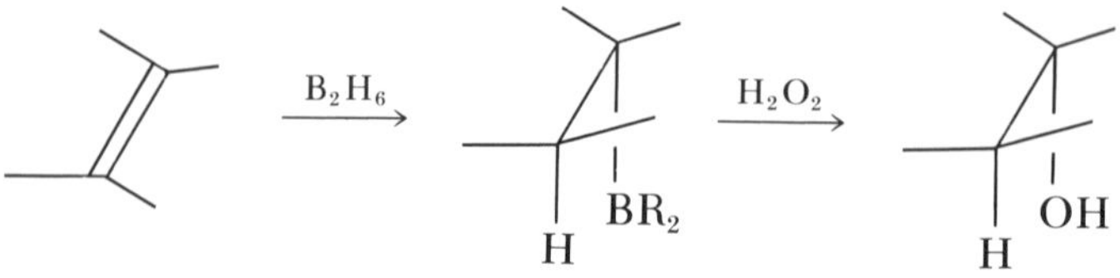

This procedure is reported to give exclusively one alcohol with norbornylene. [H. C. Brown, and G. Zweifel, *J. Am. Chem. Soc.*, **83**, 2544 (1961)] From the outcome of the borohydride reduction of norcamphor, which norborneol isomer would you predict from the B_2H_6—H_2O_2 reaction with norbornylene?

Reference

P. D. Bartlett (Ed.), Nonclassical Ions: Reprints and Commentary, W. A. Benjamin. New York, 1965.

CHAPTER 20

CHROMIC ACID OXIDATION

d 0.96   bp 156°

mp 88-90°

Chromic acid is one of the most widely used oxidizing agents in organic chemistry. A number of procedures for generating chromic acid have been developed for various purposes; among the most widely used are $Na_2Cr_2O_7$ and CrO_3 in aqueous sulfuric or acetic acid.

By proper choice of reagent and conditions, the oxidation of a secondary hydroxyl group can be carried out in a highly selective way, leaving unaffected other oxidizable groups in a molecule. On the other hand, vigorous treatment with hot chromic acid is capable of oxidizing all of the carbon atoms in a compound to the simple fragments CO_2 and CH_3CO_2H.

A mole of acetic acid is produced from any CH_3—$\overset{|}{C}$— or $(CH_3)_2\overset{|}{C}$— unit,

and this method was, prior to the advent of nmr spectroscopy, a common analytical procedure (Kuhn-Roth C-methyl determination). It is obvious, therefore, that control of the reaction conditions, particularly temperature, is quite important in a chromic acid oxidation.

Another consideration in an oxidation is the amount of oxidant required by the stoichiometry of the reaction. This is readily calculated by recognizing that for *any* C—H or C—C bond that is broken in the oxidation, the molecule *loses* two electrons. Thus the oxidation of ethylbenzene to benzoic acid and carbon dioxide requires the transfer of 12 electrons to 4 moles of CrO_3 [$Cr^{VI} \rightarrow Cr^{III}$]. The equation is then completed by adding acid to balance the charges, and water to balance the number of oxygen atoms:

$$C_6H_5CH_2CH_3 + 4\,CrO_3 + 12\,H^+ \longrightarrow C_6H_5CO_2H + CO_2 + 4\,Cr^{3+} + 8\,H_2O$$

Chromic anhydride in acetone can be used as a diagnostic tool for the qualitative detection of primary and secondary hydroxyl groups, and is complementary to the Lucas reagent (p. 125) in chemical classification of alcohols. The test is based on the characteristic color of the reduced Cr^{III} ion, and it is quite specific for $R—CH_2OH$ and $R—CHOH—R$; tertiary alcohols, amines, and olefinic double bonds generally do not react under the conditions specified.

Two procedures involving the oxidation of secondary alcohols are given below. The reactions are rather exothermic, and the experiment with cyclohexanol will provide experience in keeping a reaction under control by the proper rate of addition. If a larger run were carried out, mechanical stirring would be needed.

The oxidation of norborneol employs acetone as the solvent (Jones method). This procedure is particularly convenient for small-scale oxidations when an accurately measured amount of Cr^{VI} is required. The starting material for this preparation can be the norborneol prepared by hydration of norbornylene (Chapt. 19) or a sample of mixed norborneol isomers supplied by your instructor.

EXPERIMENTS

A. Preparation of Cyclohexanone

Procedure. In a 250 ml Erlenmeyer flask place 60 g of ice and then add 20 ml of concentrated sulfuric acid (96%). Add 20 g (20.0 ml) of cyclohexanol and mix thoroughly. (cool on)

Prepare a solution of 21 g of $Na_2Cr_2O_7 \cdot 2H_2O$ in 10 ml of water and place it in a dropping funnel mounted above the alcohol solution. With a thermometer in the alcohol-acid mixture, add the dichromate solution in small portions with swirling, keeping the temperature between 27 and 35°. The rate of addition is limited by the temperature. The reaction can be cooled by swirling in a cold-water bath to permit somewhat faster addition. Excessive cooling should be avoided, however, to prevent the possibility of building up an excess of dichromate which can then react too rapidly. When all of the dichromate has been added, and the temperature has dropped, add two ml of methanol to reduce any excess dichromate.

At this point, the cyclohexanone can be separated from the oxidation mixture and recovered either by extraction with ether and washing the ether solution, or by simple steam distillation. The latter method is somewhat more convenient. Transfer the dark green mixture to a 500-ml round-bottom flask and rinse it with about 50 ml of water. Set up the flask with a take-off head for simple distillation. Introduction of steam is unnecessary; sufficient water is present to complete the codistillation.

Distill the mixture until no more cyclohexanone droplets are present in the condensate. Add about 15 g of salt to the distillate to facilitate layer separation and depress the solubility of the ketone. Recover the ketone layer with a separatory funnel and dry with $MgSO_4$. Filter into a 50 ml distilling flask. This material should be quite pure, and the yield could reasonably be calculated at this point, but distillation is desirable. Use a simple head and collect material with a 5° boiling range in a tared receiver. Retain the distilled ketone for further use, or submit the sample to your instructor as directed.

B. Preparation of Norcamphor

The following procedure is based on a 0.01 mole scale; if a smaller amount of alcohol is used, reduce the amount of CrO_3 solution and solvents proportionately. A semimicro preparation such as this, of a low-molecular-weight, relatively volatile compound, places a premium on thoughtful experimentation and good technique. All glassware should be *clean* and *dry*. Extraction must be thorough, but if an excessive volume of solvent is used, losses will be incurred during evaporation. Volumes of wash water should be kept to a minimum, and separations should be sharp.

In a 25×100 mm test tube dissolve 1.12 g (0.01 mole) of norborneol in 8 ml of reagent grade acetone and place the solution in an ice bath. Measure (graduated pipet) 2.50 ml of $2.68M$ CrO_3 in sulfuric acid into a 10×75 mm test tube. Add the CrO_3 solution dropwise (transfer pipet and bulb) to the norborneol, shaking and swirling it in an ice bath. As soon as addition is complete, dilute the reaction mixture with 8 ml of pentane, stopper the test tube (rubber stopper), shake, and pour off the upper layer into a small separatory funnel. Extract the aqueous residue in the same way with three more 4-ml portions of pentane. (Extracting the solution in the test tube and decanting will help avoid losses due to transferring the mixture in and out of the separatory funnel.) Wash the combined pentane solutions with three or four 1 to 2-ml portions of water, adding each wash to the original aqueous phase.

Extract the combined aqueous phases once more in the test tube with 6 ml of pentane, decant the pentane as completely as possible into the funnel containing the initial pentane solutions, separate any droplets of the lower phase, wash the pentane with 3 ml of $NaHCO_3$ solution, and transfer the pentane solution to an Erlenmeyer flask, rinsing the funnel with a minimum volume of pentane. Dry with 1 to 2 g of $MgSO_4$ (only a small amount of drying agent is needed, since pentane dissolves little water) and filter the dried solution through a small plug of cotton into a dry 125 ml side arm flask. Rinse the drying agent with a few milliliters of pentane. Evaporate the solution on the steam bath (Fig. 3.6) to a thin oil, then remove, tip the warm flask slightly, and rotate it to spread the oil over the bottom and lower part of the sides as the residual pentane vapor flows out. Cool the flask to room temperature, stopper, connect to the aspirator, and evacuate *briefly* until the residue forms a solid crust. Release the vacuum immediately, insert an ice cooled cold finger (wipe off any condensed water) and sublime the product. Record the weight and melting point (sealed capillary).

C. Chromic Acid Test for Alcohols

In labeled 10×75 mm test tubes, obtain samples of 1 drop of each of the following compounds: (a) 1-butanol, (b) 2-butanol, (c) *t*-butyl alcohol, (d) benzaldehyde, (e) cyclohexene, (f) ether, (g), and (h) unknowns as directed by your instructor. Add 1 ml of acetone to each test tube.

Obtain about 2 ml of the chromic anhydride-sulfuric acid solution. To each of the acetone solutions add 1 drop of this reagent and observe and record any changes that occur within 5 seconds; regard as a negative test any reactions that occur more slowly. State whether the unknowns contained a primary or secondary hydroxyl group.

QUESTIONS

*1. Complete and balance the equations for the following reactions:

a. $CH_3CH_2CHOHCH_3 + Na_2Cr_2O_7 + H_2SO_4 \longrightarrow$
$CH_3CH_2COCH_3 + Cr^{+3}$

b. $C_6H_5CH_2OH + CrO_3 \longrightarrow C_6H_5CO_2H + Cr^{+3}$

c. $CH_3CH_2C{=}CH_2 + H_2Cr_2O_7 \longrightarrow$
$\quad\quad | \quad\quad\quad\quad\quad\quad CH_3CH_2COCH_3 + CO_2 + Cr^{+3}$
$\quad CH_3$

2. Calculate the number of moles of $Na_2Cr_2O_7$ required in the cyclohexanol oxidation; how much excess is actually used?

3. In an oxidation of cyclohexanol in which the temperature was allowed to rise to 50°, the yield of cyclohexanone was low, and an acidic byproduct, mp 153–154°, was isolated. Suggest the structure of this product.

FREQUENCY (CM⁻¹)

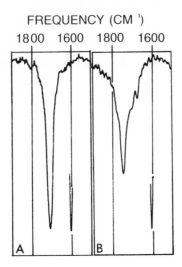

FIGURE 20.1

4. In Figure 20.1 are shown the carbonyl stretching absorptions of cyclohexanone and norcamphor. State which ketone gives which spectrum, and explain the difference in absorption frequency.

CHAPTER 21

EQUILIBRIUM CONSTANT FOR ESTERIFICATION: *n*-BUTYL ACETATE

$$CH_3CO_2H + CH_3CH_2CH_2CH_2OH \overset{H^+}{\rightleftharpoons} CH_3CO_2CH_2CH_2CH_2CH_3 + H_2O$$

Direct acid-catalyzed esterification is a classic method for preparing esters, but it is a reversible reaction, and an equilibrium is established. For preparative purposes, water is removed, usually by codistillation with an immiscible solvent, driving the reaction to the right.

The equilibrium constant is expressed by the equation

$$K = \frac{(C_{ester})(C_{H_2O})}{(C_{acid})(C_{alcohol})} \tag{21.1}$$

If equimolar amounts of acid and alcohol are used, and we assume no volume change (permitting use of molar amounts directly instead of concentrations in Eq. 21.1), the equilibrium constant K can be determined simply by titration of the amount of acid present at the beginning and at equilibrium. The amount of acid can be obtained directly from the volume of base consumed by an aliquot; the base used in the titration is not accurately standardized since the exact normality is not required.

Let n = moles acid = moles alcohol initially = $V_o N$ of base

   x = moles acid = moles alcohol at equilibrium = $V_E N$ of base

 $n - x$ = moles ester = moles water at equilibrium

and $K = \dfrac{(n-x)(n-x)}{(x)(x)} = \dfrac{(n-x)^2}{x^2}$

In this experiment, the equilibrium constant for the reaction of *n*-butanol and acetic acid is to be determined, and the ester will then be isolated.

EXPERIMENTAL PROCEDURE

Equilibrium Constant

In a 100 ml boiling flask place 25.9 g (0.35 mole) of *n*-butanol and 21 g (0.35 mole) of glacial acetic acid. After mixing thoroughly (swirling), transfer a 1.00 ml sample of the solution to a 125 ml Erlenmeyer flask, containing 20 ml of water and a few drops of phenolphthalein solution, and titrate to a pink endpoint with 0.5N sodium hydroxide to obtain V_o.

Add *4 drops* of concentrated sulfuric acid (this is about 100 mg) to the alcohol-acid mixture, add a boiling stone, and reflux for 40 minutes. A very small upper layer of ester will separate during this reflux period. Cool the solution and remove another 1.00 ml sample; be sure to take this sample from the *lower* layer (insert pipette below surface, expel a few bubbles of air, and then fill pipette). Titrate this sample as before, with the *same* NaOH solution.

Reflux the solution (fresh boiling stone) for another 20 minutes, and again cool, remove a 1.00 ml aliquot from the lower layer and titrate. If the volume of base in this titration deviates more than 0.3 ml from that in the previous one, repeat the reflux for another 30 minutes and again titrate a 1.00 ml sample. The final titration gives V_E, representing acid present at equilibrium.

From these data, calculate the equilibrium constant for the esterification, neglecting the small error that may be introduced by the separation of two phases. (Assume volume remains constant throughout.)

Isolation of *n*-Butyl Acetate

Place the ester solution under a fractionating column with condenser, adapter, and thermometer and distill until about 20 ml of distillate has been collected. Separate the two layers in the distillate, record volume of aqueous (lower) phase, and return the upper phase to the distilling flask. This upper layer contains mainly ester and unreacted alcohol.

Repeat the distillation, collecting 15 to 20 ml, and again

return the upper phase to the flask and record the volume of the lower phase. If an appreciable volume of water is obtained in this second distillation, repeat the distillation again.

The total organic material, in flask and distillate, is then combined and diluted with an equal volume of ether to reduce mechanical losses (rinse out the column and condenser with ether and add this to main solution). The ether solution is washed with aqueous carbonate until the wash is not acidic. Dry the ether solution over $MgSO_4$, filter through a cotton plug into a 100 ml round-bottom flask, and distill. A fractionating column is not needed, but distillation should be carried out slowly when the temperature begins to rise. Collect the ester over a 5° range (lit. bp 126°) and determine the yield.

QUESTIONS

1. The apparatus shown in Figure 21.1 is a Dean-Stark trap, designed for removing water from a reaction mixture containing a water-immiscible solvent. Sketch an equipment set-up for the preparation of 500 g of n-butyl acetate, incorporating this type of trap into the apparatus, and show how it would function to permit removal of water during the reaction.

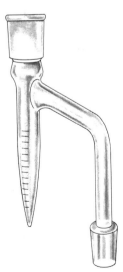

FIGURE 21.1 Dean-Stark trap.

2. Why is the normality of the aqueous base not required in the titration of the reaction aliquots?

3. Assume that the amount of H_2SO_4 added as a catalyst is exactly 100 mg. Calculate the error that was introduced in the K_{eq} determination by neglecting to correct for this added acid.

A meticulous student decided to avoid this error by titrating the solution to obtain V_0 both before and after adding the H_2SO_4. He added the sulfuric acid, swirled the solution for a minute, and then removed a 1-ml aliquot and titrated it. This titration required *less* NaOH than did the aliquot removed prior to adding the H_2SO_4. What explanation could you offer?

EFFECT OF SUBSTITUENTS ON THE ACIDITY OF BENZOIC ACID

The purpose of this experiment is to measure the effect of various ring substituents on the acidity of benzoic acid. Acidity is defined as the degree of dissociation of an acid into a carboxylate anion and a hydronium ion (Eq. 22.1), and it is measured in terms of an equilibrium constant, K_a, which is a

$$\text{(22.1)}$$

HA $\qquad\qquad$ A$^-$ $\qquad$ H$^+$

function of the concentration of these species at equilibrium (Eq. 22.2). Since it is relatively easy to measure $[H^+]$, one can calculate K_a if the relative

$$K_a = \frac{[H^+][A^-]}{[HA]} = [H^+]\frac{[A^-]}{[HA]} \qquad (22.2)$$

concentrations of A$^-$ and HA are known. In fact if $[A^-] = [HA]$, $K_a = [H^+]$. One way of establishing this situation would be to dissolve an equal molar mixture of the acid and its sodium salt in water. However, it is simpler to start with a solution of the acid and titrate it with a strong base until it is half-neutralized. This latter method will be used in this experiment.

Since devices for determining acid concentration do not give $[H^+]$ directly, but rather pH $(= -\log[H^+])$, the acidity can be expressed as pK_a (Eq. 22.3).

149

$$pK_a = -\log K_a$$
$$= -\log [H^+] - \log([A^-]/[HA])$$
$$= pH - \log([A^-]/[HA]) \tag{22.3}$$

when $[A^-] = [HA]$

$$pK_a = pH - \log(1) = pH$$

EXPERIMENTAL PROCEDURE

From the sample of the substituted benzoic acid provided, weigh out approximately 20 mg and dissolve it in 25 ml of ethanol in a 100 ml beaker. Dilute the solution with 25 ml of water, and titrate with $0.01N$ sodium hydroxide solution. Your instructor will explain how to use the pH meter. The titrant should be added from a burette in 1.0-ml increments, the solution mixed well, and the pH recorded in a table of pH vs total volume of titrant added. Near the endpoint, i.e., where the pH changes most rapidly, the pH can be measured between 0.5 ml additions to aid in plotting the results later. Titrate 3 to 4 ml beyond the endpoint.

Plot the measured pH values vs volume of titrant on an appropriately scaled sheet of graph paper (Fig. 22.1). Draw a smooth curve through the points and mark the point at which the curve is steepest. From the graph obtain the exact volume (V_1) corresponding to this point. At one-half this volume

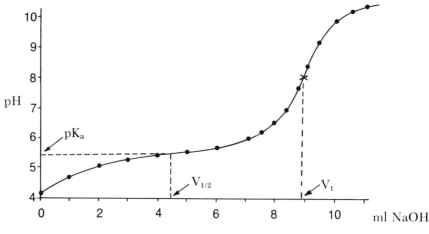

FIGURE 22.1 pK$_a$ titration curve.

($V_{1/2}$), one half of the acid is neutralized, i.e., $[A^-] = [HA]$, and the pH at this point equals the pK_a of the acid. Replot the points around $V_{1/2}$, expanding the scale of the pH axis, to permit a more accurate measurement of the pK_a.

Report the pK_a of your acid to your instructor who will tabulate the results obtained on all the acids given out. Order this list of acids according to their acidity (*note:* higher acidity= larger K_a = smaller pK_a). Explain the order and any differences between the effect of *meta-* and *para-*substitution.

***

Reference

J. F. King, In *Techniques of Organic Chemistry*, Vol. XI, Part 1, K. W. Bently (Ed.), Interscience, New York, 1963, pp. 318–369.

23

EFFECT OF SUBSTITUENTS
ON REACTIVITY

In the preceding experiment the effect of various ring substituents on the acidity of benzoic acids was observed. Since the dissociation constant of an acid (K_a) is the ratio of the rate of dissociation of the acid divided by the rate of protonation of the carboxylate anion (Eq. 23.1), it follows that these rates are affected by ring substituents.

$$(23.1)$$

$$K_a = \frac{k_1}{k_{-1}}, \qquad pK_a = -\log K_a$$

Rates of other reactions of groups attached to benzene rings are also affected by substituents on the ring. In this experiment the rates of base-catalyzed hydrolysis (saponification) of several substituted methyl benzoates will be measured (Eq. 23.2). As in the preceding experiment, different esters will be used by different students, and the results of all will be compared.

$$(23.2)$$

EXPERIMENTS

Preparation of Methyl Benzoates

Place 5 g of the acid provided in a 100 ml round-bottom flask and add 50 ml of methanol and approximately 1 ml (one dropperful) of concentrated sulfuric acid. Fit a reflux condenser to the flask, add a boiling stone, and reflux the mixture for 1 to 2 hours. The progress of the reaction can be checked at various times (every 20 minutes or so) by TLC on silica gel. The solvent or solvent mixture required for this will depend on the acid being used and will have to be determined individually. Before taking the first sample, find a solvent which barely, but perceptibly, moves the acid from the origin. The ester, being less polar, will have a larger R_F and will move well up the plate.

When the reaction is complete as measured by TLC analysis, cool the solution to room temperature, and transfer it to a 250 ml separatory funnel. Add 50 ml of ethyl ether and 75 ml of water and shake well. Allow the layers to separate and discard the aqueous layer. Rinse the organic layer successively with 25 ml of water, twice with 25 ml of 10% $NaHCO_3$ solution and once with 25 ml of saturated NaCl solution. Further dry the organic layer over Na_2SO_4 for 5 to 10 minutes, filter out the drying agent, rinse it with ether, and evaporate the solvent from the filtrates.

Most of the esters assigned can be obtained as solids and purified by recrystallization from hexane or a mixture of benzene and hexane (exceptions: methyl benzoate, methyl *m*-toluate, methyl *m*-chlorobenzoate, and methyl *m*-anisate).

Some of the solid esters are relatively low melting, so that their recrystallization must be done at ice bath temperatures. For use in the following experiment, the esters must be completely dried of the recrystallization solvent. Those which are oils should be distilled from a small distilling flask. Due to their high boiling points (190 to 260°) a condenser will not be needed; the set-up in Figure 35.1 will suffice. Report the observed and literature melting (or boiling) range of the ester and the per cent yield.

Rate Measurements

Weigh out 0.0100 mole of the methyl ester prepared in the preceding section into a tared 125 ml Erlenmeyer flask; the weight should be measured accurately and should differ from 0.0100 mole by no more than 1 per cent. Weigh 60.0 ml of reagent acetone (d .791) into the flask and swirl the mixture to dissolve the ester completely. Suspend the flask, using a buret clamp in an ice-water bath. During the reaction ice should be added as necessary to keep the temperature at 0°.

Weigh out 40.0 ml of 0.250 M NaOH (d = 1.011) solution and cool to 0°. To start the reaction add this to the ester solution and mix well by stirring. Note the time, using a watch or clock with a second hand, when the base is added. With a 10 ml pipette remove 10 ml aliquots of the solution at regular intervals.* Add each sample to a separate flask containing 10 ml (measured by pipette) of 0.100N HCl, and note the time when the solution is added.

Add 3 drops of phenolphthalein indicator solution to each flask and titrate each with 0.010N NaOH solution to a persistent pink endpoint. Record for each sample the time when it was taken and the volume of titrant required. The acid being titrated is the benzoic acid liberated in the saponification.† Using the volume and normality of titrant, calculate the number of moles of acid in each sample and from this the number of moles of unreacted ester remaining. The concentration of ester can be calculated by simply dividing the number of moles by the sample volume. All of the preceding data for each sample should be listed in a table similar to Table 23.1. The seventh column, the reciprocal of the sixth, will be used in the calculation of k.

TABLE 23.1

SAMPLE NUMBER	TIME TAKEN(x)	VOL. OF .01N NaOH	mmoles OF ACID	mmoles OF ESTER	CONCEN-TRATION OF ESTER	(CONCEN-TRATION OF ESTER)$^{-1}$ (y)
1	65 sec	5.1 ml	.051	.95	.095 M	10.5 M^{-1}
2	250 sec	9.9 ml	.099	.90	.090 M	11.1 M^{-1}
3	600 sec	35.3 ml	.353	.647	.065 M	15.4 M^{-1}
4	1200 sec	62.1 ml	.621	.349	.0379 M	26.4 M^{-1}
5	1800 sec	76.6 ml	.766	.234	.0234 M	42.7 M^{-1}

Calculation of the Rate Constant for Saponification

Since the slow (rate determining) step in the reaction of Eq. 23.2 is the attack of OH$^-$ on the ester (Eq. 23.3), the rate of

---

*The optimum length of the intervals will depend on which ester you are using. See your instructor for directions on this point.

†This is so because after x per cent of the ester has reacted there are in solution $0.01\left(1-\dfrac{x}{100}\right)$ mole each of ester and OH$^-$, and $0.01\left(\dfrac{x}{100}\right)$ mole of benzoate anion. Addition of 10 ml of 0.100N H$^+$ gives $0.01\left(1-\dfrac{x}{100}\right)$ moles of H$_2$O (from OH$^-$) plus $0.01\left(\dfrac{x}{100}\right)$ moles of the benzoic acid.

disappearance of ester is proportional to the product of the

$$R-C{\overset{\displaystyle O}{\Big\diagdown}}_{OMe} + {}^-OH \overset{slow}{\longrightarrow} R-C{\overset{\displaystyle O^-}{\underset{\displaystyle OH}{\Big|}}}_{OMe} \overset{fast}{\longrightarrow} R-C{\overset{\displaystyle O}{\Big\diagdown}}_{O^-} + MeOH \quad (23.3)$$

concentrations of these two species (Eq. 23.4).

$$\text{rate} = -\frac{d[\text{ester}]}{dt} = k[\text{OH}^-][\text{ester}] \qquad (23.4)$$

In this experiment $[\text{OH}^-] = [\text{ester}]$, so that it is possible to simplify Eq. 23.4 to Eq. 23.5.

$$-\frac{d[\text{ester}]}{dt} = k[\text{ester}]^2 \qquad (23.5)$$

After integration of this differential form of the rate equation, one obtains Eq. 23.6, where $[\text{ester}]_t$ is the concentration of ester after t seconds of reaction.

$$\frac{1}{[\text{ester}]_t} = kt + \frac{1}{[\text{ester}]_o} \qquad (23.6)$$

The value of k can be obtained by substituting values of $[\text{ester}]_t$, t, and $[\text{ester}]_o$ (=0.1M) into Eq. 23.6. However, it is simpler and more accurate to treat Eq. 23.6 as that of a straight line plot of $[\text{ester}]_t^{-1}$ vs t, whose slope is k. Therefore, to obtain k for your ester, plot the values of $[\text{ester}]_t^{-1}$ (column 7 in Table 23.1) vs time (column 2) on graph paper with appropriate scales, and then with a straightedge draw a line which passes as close as possible to all of the points. (Due to experimental errors, you should not expect that the line will go through each of the points.) Using two well-separated points on the line (not necessarily data points), (x_1, y_1) and (x_2, y_2), calculate the slope of the line, $k = (y_2 - y_1)/(x_2 - x_1)$. Since the ordinate (y) is in units of M^{-1} (liters/mole) and the abscissa (x) is in units of sec, the rate constant has units of M^{-1} sec^{-1}.

After everyone in the class has calculated a value of k for a different ester, the results can be collected and compared. For this purpose it is useful to prepare a table of esters and rate constants in order of increasing values of k. How does this list compare with the corresponding list of acids and pK_a's from the previous experiment? Why? If a more quantitative comparison of the two sets of quantities seems justified, you can make a graph of the values of log k for the esters *vs* the pK_a's ($-\log K_a$'s) of the corresponding acids.

Reference

A. A. Frost, and R. G. Pearson, *Kinetics and Mechanism.* 2nd Ed. John Wiley and Sons, New York, 1961, pp. 8–24.

CHAPTER 24

ANALYSIS OF FATS
AND OILS

The bulk of plant and animal tissue is composed of three main classes of compounds: carbohydrates, proteins, and lipids. Lipids consist largely of fats, which are triglycerides, i.e., carboxylic esters of glycerol.

$$
\begin{array}{ccc}
\overset{O}{\underset{}{R_1 C}}\!\!-O-CH_2 & R_1 CO_2^- & HOCH_2 \\[1em]
\overset{O}{\underset{}{R_2 C}}\!\!-O-CH & \xrightarrow{\ OH^-\ } R_2 CO_2^- + HOCH \\[1em]
\overset{O}{\underset{}{R_3 C}}\!\!-O-CH_2 & R_3 CO_2^- & HOCH_2 \\[0.5em]
\text{Triglyceride} & \text{Glycerol}
\end{array}
\qquad (24.1)
$$

The principal acids in animal fats and vegetable oils are of the type $CH_3(CH_2)_n(CH_2 CH{=}CH)_m(CH_2)_7 CO_2 H$, where $m = 0-3$ and $n = 1-13$. In addition, some of the shorter straight-chain acids (butyric-octanoic) are occasionally present. The names of these acids are derived from the plant or animal source; the acids present in a number of common fats and oils are listed in Table 24.1.

It is possible to determine the source of a fat or oil by analysis of the component acids, since the composition is quite characteristic for a given oil. The simplest procedure for this analysis is conversion of the acids to their methyl esters and determination of the relative amounts of each by VPC.

The classical procedure for obtaining the acids from a fat is saponification with alkali (Eq. 24.1) to give the salt (a soap), and then acidification. This hydrolysis is rather slow at room temperature, and at high temperatures, when a glycol is used as solvent, isomerization of the polyunsaturated acids occurs. For analytical purposes, where only a small amount of ma-

TABLE 24.1 *TYPICAL FATTY ACID CONTENT OF SELECTED FATS AND OILS*

	CONSTITUENT FATTY ACIDS (%)[a]								
No. Carbon Atoms: <12	12	14	16	16	18	18	18	18	>18
No. Double Bonds:	0	0	0	1	0	1	2	3	
Acid Name	lauric	my-ristic	pal-mitic	palmi-toleic	steric	oleic	lino-leic	lino-lenic	
Human Fat —	—	3	24	5	8	47	10	—	2
Butterfat 4[b]	4	12	33	2	12	29	2	—	2
Lard —	—	1	28	1	16	43	9	—	2
Coconut Oil 15[c]	46	18	10	—	4	6	—	—	—
Corn Oil —	—	—	10	1	2	25	62	—	—
Cottonseed Oil —	—	1	20	2	1	20	55	—	1
Linseed Oil —	—	—	10	—	4	23	57	6	—
Olive Oil —	—	—	9	—	3	78	10	—	—
Palm Oil —	—	1	45	—	4	40	10	—	—
Peanut Oil —	—	—	10	—	3	48	34	—	5[d]
Safflower Oil —	—	—	7	—	2	13	75	3	—
Soybean Oil —	—	—	9	—	4	43	40	4	—

[a]Blank spaces indicate that less than 1 per cent of the corresponding acid is usually present.

[b]C_6 saturated (caproic acid) and C_{10} saturated (capric acid).

[c]C_8 saturated (caprylic acid) and C_{10} saturated.

[d]C_{20} saturated (arachidic acid), C_{22} saturated (behenic acid), and C_{24} saturated (lignoceric acid).

terial is required, this difficulty is avoided by preparing the methyl esters directly from the triglyceride by transesterification (Eq. 24.2). By using a large excess of methanol, the equilibrium is shifted essentially completely to the right.

$$\underset{\text{R—C—OR}'}{\overset{\text{O}}{\|}} + \text{MeOH} \xrightleftharpoons{\text{acid}} \underset{\text{R—C—OMe}}{\overset{\text{O}}{\|}} + \text{R}'\text{OH} \qquad (24.2)$$

In this experiment, you will obtain a sample of a commercially available fat or oil (e.g., a cooking oil) to identify by comparison of its fatty acid content with the compositions of various oils listed in Table 24.1. These data are typical, but vary by ±5 per cent among samples of the same oil from different sources.

EXPERIMENTAL PROCEDURES

Transesterification of triglycerides

Weigh 0.10 g (approximately 8 drops) of the fat or oil into a 1 dram (4 ml) screw cap vial. Add 2 ml of benzene to dissolve, and 1 ml of the solution of BF_3 in methanol (12 g BF_3/100 ml MeOH). Screw on the cap tightly, shake to mix, and place in a 50 ml beaker, half full of boiling water. Check to see that no vapor escapes; if hissing or bubbling is observed, cool and re-tighten the cap. After 15 minutes, remove the vial, cool, and add 1 ml of water. Shake well and let the mixture stand to separate (10 to 15 minutes). The benzene (upper) layer contains the methyl esters.

VPC analysis of methyl esters

Inject 2 μl of the benzene solution onto a polar column (diethyleneglycol succinate liquid phase) in the VPC. By comparison with a VPC of a standard mixture of fatty acid esters (provided), identify the components and calculate the approximate per cent composition of the mixture. Relative areas may be measured by tracing the peaks onto bond paper using carbon paper, cutting out the tracings, and weighing them ($\pm$.01 g) on a balance (see question 3). Compare the composition with those in Table 24.1 and identify the original oil.

QUESTIONS

⋆1. What is the purpose of the BF_3 in the transesterification reaction? Give equations.

2. What happened to the glycerol from the triglyceride; i.e., why was it not observed in the VPC?

3. An alternative to cutting and weighing the peaks to determine their areas is the following procedure. Multiply the peak heights by their retention times, sum the products, and divide each product by the total. Suggest why this procedure works for this mixture of similar compounds (see p. 48), and if time permits, compare the results by both methods.

4. Assuming an essentially quantitative transesterification, estimate what amount (in grams) of ester corresponds to the smallest detectable peak in your VPC tracing (i.e., the sensitivity of the detector).

References

W. R. Morrison, and L. M. Smith, *J. Lipid Res.*, **5**, 600 (1964).

P. L. Nichols, S. F. Herb, and R. W. Riemenschneider, *J. Am. Chem. Soc.*, **73**, 247 (1951).

CHAPTER 25

PHOTOCHEMICAL REACTIONS

Photochemical reactions, involving molecules in high energy excited states, provide synthetic pathways to otherwise inaccessible compounds. A simple illustration is the photoisomerization of 1,3-cyclohexadiene to bicyclo[2.2.0]hexene:

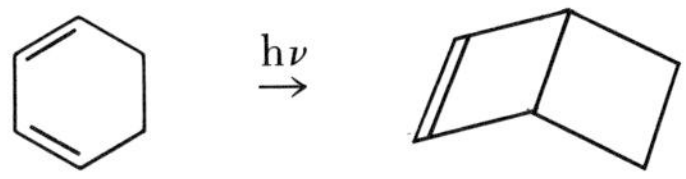

Other types of photochemical reactions include dimerizations (cycloadditions) and oxidation-reduction processes. Examples of these are given in the following two experiments.

A. PHOTOCHEMICAL DIMERIZATION OF *TRANS*-CINNAMIC ACID

This reaction is a classic example of the photochemical cycloaddition of two alkenes to form a cyclobutane. For unsymmetrically substituted alkenes such as cinnamic acid (**1**), two modes of reaction are possible, leading to head-to-head (**2**) or head-to-tail (**3**) dimers.

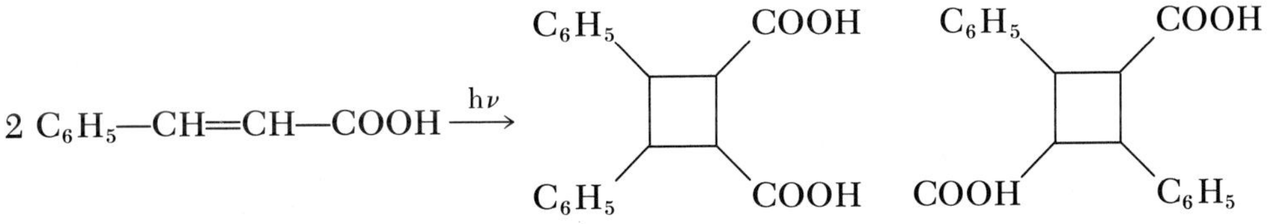

1 2 3

Truxinic acids Truxillic acids

The stereochemistry of the product poses a complex problem since there are six diastereomers of the truxinic acids and five of the truxillic acids. The photodimerization of cinnamic acid is carried out with the crystalline solid, and the product depends on the particular crystalline form that is irradiated. This experiment illustrates how one might elucidate the structure of a given isomer using spectral and chemical information.

The problem was first solved long before the discovery of nmr spectroscopy and required a much more elaborate chemical analysis. It did use the knowledge that only certain of the isomeric diacids readily form cyclic anhydrides (4) or (5), indicating that in these isomers the carboxyl groups are *cis* to one another. Other isomers, which do not give cyclic anhydrides under the normal conditions (e.g., heating with acetic anhydride) are transformed to anhydrides when sodium acetate is added to the reaction mixture. In this case, sodium acetate functions as a base to invert the configuration (epimerize) at one of the carboxyl-substituted carbons, giving a stereoisomer which can form a cyclic anhydride.

4   5

In this experiment, *trans*-cinnamic acid will be photodimerized and the dimeric acid converted to derivatives whose nmr spectra permit assignment of the structure.

EXPERIMENTAL PROCEDURE

Dimerization of Cinnamic Acid

Weigh out 1.5 g of *trans*-cinnamic acid and transfer it to a 125 ml Erlenmeyer flask. By heating on a steam bath, dissolve the acid in approximately 2 ml of tetrahydrofuran (THF). Remove the flask from the bath, and while the solution is still hot, rotate the flask to coat the walls with the crystallizing acid. If a very uneven coating is obtained, reheat to redissolve the acid and repeat the process. When the residue is sufficiently dry, clamp the flask in an inverted position for 30 minutes to permit all solvent vapor to flow out. Stopper the flask with a cork, label it with your name, and place it, cork down, in a beaker on a window ledge (with southern exposure if possible) for irradiation by the sunlight. After one week, rotate the flask by 180° to expose the opposite side.

After the second week of irradiation, transfer the solid to a 25×100 mm test tube, and add 20 ml of benzene to dissolve the cinnamic acid remaining. Collect the product in a Hirsch funnel, and wash it in the funnel with 20 ml more benzene. Air-dry and weigh the product, and record the per cent yield and melting point.

Proceed with one of the two following experiments as directed by your instructor.

Esterification of Cinnamic Acid Dimer

Transfer the acid to an 18×150 mm test tube and add 10 ml of methanol, 2 to 3 drops of concentrated sulfuric acid, and a boiling stone. Assemble a test tube reflux apparatus as shown in Figure 12.1 and reflux the mixture gently on a steam bath. After one hour of refluxing, cool the solution, add 25 ml each of water and ether, shake well, and separate the aqueous layer. Extract the ether layer with $NaHCO_3$ solution, and dry $(MgSO_4)$. Filter and concentrate the solution to small volume, transfer to a 25×100 mm test tube, evaporate to dryness, and recrystallize the residue from a minimum amount of hexane or methanol. Collect and air-dry the recrystallized diester and report the yield and melting point. On the basis of its melting point and that of the original diacid, which of the isomers in Table 25.1 was obtained? The nmr spectrum of the diester is given in Figure 25.1.

TABLE 25.1

ISOMER	MP OF ACID	MP OF ANHYDRIDE*	MP OF DIMETHYL ESTER	MeO-PROTON PEAKS IN NMR OF ESTER
1	286°C	—	174°C	1
2	285°	—	112°	2
3	266°	287°C	104°	1
4	245°	unknown	133°	1
5	239°	150°	116°	2
6	228°	191°	127°	1
7	210°	116°	76°	1
8	209°	—	127°	2
9	196°	—	199°	1
10	192°	unknown	64°	1
11	175°	—	77°	1

*A dash (—) indicates that the acid cannot form a cyclic anhydride; "unknown" means that the anhydride can presumably exist but has not been prepared.

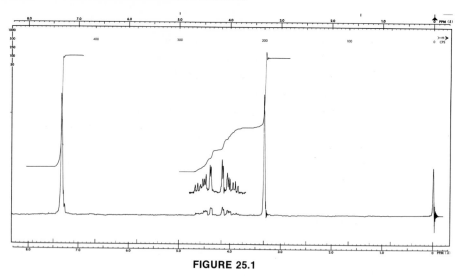

FIGURE 25.1

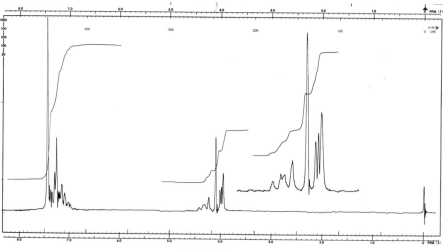

FIGURE 25.2

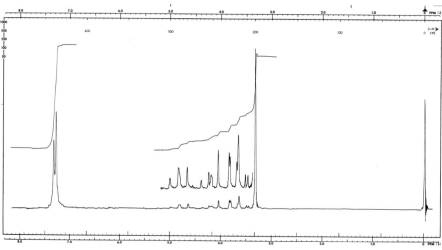

FIGURE 25.3

Epimerization and Dehydration of Cinnamic Acid Dimer

Place the acid, along with 0.3 g of sodium acetate, 2 ml of acetic anhydride, and a boiling stone, in a 15×150 mm test tube assembled as a reflux apparatus as shown in Figure 12.1. With a small flame, gently reflux the mixture for 10 minutes. Add dropwise to the warm mixture 1 ml of water, and after the exothermic hydrolysis of the acid anhydride is complete, add another 5 ml of water and transfer the mixture to a small separatory funnel. Rinse the test tube with 20 ml of water and 10 ml of chloroform, shake well, and separate the layers. After washing with $NaHCO_3$ solution, dry over $MgSO_4$, and concentrate the solution to approximately 2 ml on the steam bath. Add ethanol to the refluxing solution (2 to 4 ml) until crystals begin to form, cool, and collect the anhydride. Rinse the crystals with 2 to 3 ml of ethanol and air-dry, weigh, and determine the melting point. A second crop may be obtained by concentrating the filtrate. The nmr spectrum of this anhydride is shown in Figure 25.2.

This anhydride may be converted to the corresponding dimethyl ester using the procedure given for the esterification of the cinnamic acid dimer. The progress of the reaction can be followed conveniently by TLC on silica gel with chloroform (see question 7). The nmr spectrum of the diester formed is shown in Figure 25.3.

QUESTIONS

*1. Write out the structures of the six truxinic and five truxillic acids.

*2 . Which of the acids in question 1 can form cyclic anhydrides?

*3. Which of the acids will give methyl esters with identical methoxyl groups by nmr?

*4. Which of the truxillic and truxinic acids might be converted into each other by epimerization (inversion of configuration) at one of the carboxyl-substituted carbons?

*5. Which of the truxillic or truxinic acids is isomer no. 5 in Table 25.1?

6. Using the data in Table 25.1, and the data from your experiment, determine the structure of the compound formed by the photochemical dimerization of cinnamic acid.

7. What did you observe by TLC during the anhydride to diester conversion? Account for this by equations for the reactions.

8. Assign all the peaks in the nmr spectra (Figs. 25.1 to 25.3).

B. PHOTOCHEMICAL PREPARATION OF BENZOPINACOL

This experiment provides an exceedingly simple and efficient example of a photochemical reduction of a diaryl ketone to a pinacol, using a secondary alcohol as the hydrogen donor.

$$2 \begin{array}{c} C_6H_5 \\ \diagdown \\ C{=}O \\ \diagup \\ C_6H_5 \end{array} + R_2CH{-}OH \longrightarrow C_6H_5{-}\overset{\overset{\displaystyle OH}{|}}{\underset{\underset{\displaystyle C_6H_5}{|}}{C}}{-}\overset{\overset{\displaystyle OH}{|}}{\underset{\underset{\displaystyle C_6H_5}{|}}{C}}{-}C_6H_5 + R_2C{=}O$$

EXPERIMENTAL PROCEDURE

In an 18×150 mm test tube place 2 g of benzophenone. Dissolve the ketone in about 8 ml of isopropyl alcohol with gentle warming, add one drop of glacial acetic acid, and then fill the tube just to the top with more isopropyl alcohol, so that air is excluded during the reaction. Stopper the tube tightly with a cork of the proper size. Hold the cork in place with a strip of plastic tape. Invert the tube in a small beaker to present the maximum surface to the light, label with your name, and place on the window sill for one week. After several sunny days, the reaction should be complete, with a heavy deposit of large crystals. Collect the product, label with melting point and yield, and submit it to your instructor.

QUESTIONS

1. The photochemical pinacol reduction begins with the excitation of benzophenone, followed by abstraction of a hydro-

gen atom from 2-propanol to give two radicals. The reaction occurs less efficiently with ethanol and not at all with *t*-butyl alcohol. In the second step of the reaction, a hydrogen atom is transferred from the radical derived from 2-propanol to a second molecule of benzophenone. Combination of two radicals derived from benzophenone gives the pinacol. Write equations showing these three steps, with the overall formation of benzopinacol and acetone.

2. The acetic acid is added to suppress a subsequent reaction that can occur with benzopinacol and traces of base. This reaction involves cleavage to give one molecule of benzophenone. Write an equation for this reaction, showing the other product(s) and the mechanism.

3. When the benzopinacol is warmed with acetic acid, rearrangement occurs to give an isomeric product, benzopinacolone. Write the structure of this product and indicate the mechanism by which it is formed.

26

PREPARATION OF BENZOIC ANHYDRIDE

Traditional procedures for converting a carboxylic acid to the anhydride include refluxing the acid with acetic anhydride and reaction of the sodium salt with the acid chloride. More practical methods are available, however, which depend on the reaction of a carboxylic acid with a sulfonyl or acylpyridinium ion. The acylpyridinium salts are very readily formed from the acyl or sulfonyl chloride and are responsible for the catalytic role of pyridine in many acylation reactions.

In this experiment, benzoic anhydride is obtained by the extremely simple reaction of the acid with a sulfonylpyridinium chloride. An intermediate mixed sulfonic-carboxylic anhydride is presumably first formed, and this reacts with more acid to give the anhydride.

$$\text{(pyridine)} + C_6H_5SO_2Cl \longrightarrow \text{(N}^+\text{--SO}_2C_6H_5\text{)}$$

$$\underset{C_6H_5CO_2H}{\searrow}$$

$$\underset{O}{\overset{O}{\underset{\|}{C_6H_5C}}}\text{--O--SO}_2C_6H_5 \xrightarrow{C_6H_5CO_2H} \underset{O}{\overset{O}{\underset{\|}{C_6H_5C}}}\text{--O--}\underset{O}{\overset{O}{\underset{\|}{C}}}C_6H_5$$

EXPERIMENTAL PROCEDURE

In a 50 ml Erlenmeyer flask place 3.0 g of benzoic acid and add 5 ml of pyridine. When the acid has dissolved, add 1.6 ml (2.2 g) of benzenesulfonyl chloride. Allow the solution to stand for 5 minutes, then pour it into about 30 ml of cracked ice and

168

induce crystallization of the oil by scratching. Collect the solid benzoic anhydride by suction filtration and wash the moist filter cake on the funnel (disconnect vacuum tube) with ice-cold $NaHCO_3$ and then water, reapply suction, and press as dry as possible. Transfer the solid to a 50 ml Erlenmeyer flask and recrystallize from ether-petroleum ether. (Dissolve in the minimum volume of ether, dilute with equal volume of petroleum ether, swirl for a few minutes with just enough $MgSO_4$ to give clear solution, decant, chill, and add more petroleum ether, if necessary, until incipient turbidity, and add seed of original solid.) Collect and air-dry the anhydride; it may be desirable to collect a second crop from the mother liquor. Report yield and melting point.

QUESTION

1. In another preparative method for this anhydride, using this same general approach, benzoyl chloride and excess pyridine are combined, and one-half molecular equivalent of water is added. Write equations for the reactions in this case, and calculate the amounts of benzoyl chloride, pyridine, and water necessary to obtain the same amount of anhydride that you obtained in this experiment, assuming an overall yield (isolated) of 80 per cent.

Reference

J. H. Brewster and C. J. Ciotti, Jr., *J. Amer. Chem.*, **77**, 6214 (1955).

CHAPTER 27

BENZYNE: PREPARATION OF 1,4-DIHYDRONAPHTHALENE-1,4-ENDOXIDE

The molecule benzyne (**1**) was first recognized as an intermediate in certain nucleophilic substitution reactions in 1953, and has since been generated by several different procedures. The main synthetic utility of

benzynes arises from their great reactivity as electrophiles and dienophiles. As examples of the former property, when generated in water or liquid ammonia, benzyne rapidly reacts to form phenol or aniline, respectively. The following experiment illustrates its Diels-Alder reactivity, in which furan is the diene. The 1,4-dihydronaphthalene-1,4-endoxide (**2**) produced is readily isomerized to 1-naphthol by acid.

170

EXPERIMENTAL PROCEDURE

1,4-Dihydronaphthalene-1,4-endoxide

In a 100 ml round-bottom flask place 10 ml of furan (d. 0.94), 10 ml of 1,2-dimethoxyethane (glyme), and a boiling stone, and attach a reflux condenser. In separate 25×100 mm test tubes, dissolve 4 ml of *iso*-amyl nitrite (d. 0.87) in 10 ml of glyme and 2.75 g of anthranilic acid in 10 ml of glyme. Heat the furan solution to reflux on a steam bath and at 3- to 4-minute intervals add 1 ml of each of the other two solutions through the condenser. Use separate transfer pipettes for each of the two reagents since they react with one another. After the additions are complete continue refluxing for 5 minutes.

Prepare a solution of 0.5 g (5 pellets) of NaOH in 25 ml of water, and add this to the cooled reaction mixture. Transfer the mixture to a separatory funnel and extract with 25 ml of pentane. Discard the aqueous layer and rinse the pentane solution with six 15-ml portions of water. Add $MgSO_4$ and charcoal to dry and partially decolorize the pentane solution, filter, and concentrate to approximately 10 ml on a steam bath. If an oil separates at this point, decant the warm pentane solution into a clean test tube and rinse the oil with 1 to 2 ml of pentane. Let the solution cool and scratch to induce crystallization of the endoxide. Collect the product and recrystallize it from about 5 ml of pentane. Weigh, determine the melting point, and calculate the per cent yield of 1,4-dihydronaphthalene-1,4-endoxide, (lit. mp 56°). A purer product can be obtained by sublimation at 100° under aspirator vacuum (see Chapt. 19) rather than recrystallization, if time permits.

1-Naphthol

Place 0.50 g of 1,4-dihydronaphthalene-1,4-endoxide and 10 ml of ethanol in a 25 × 100 mm test tube, and add 5 ml of concentrated hydrochloric acid. Stir and let the mixture stand for 10 minutes. Transfer the mixture to a separatory funnel, rinsing the test tube with 20 ml of ether and 15 ml of water. Shake and separate the layers. Rinse the ether layer with two 5-ml portions of water, dry over Na_2SO_4, and evaporate to an oil on the steam bath in a 25 × 100 mm test tube. Add 15 ml of hexane, heat to dissolve the product, and let cool to recrystallize. Filter out the slightly pink crystals, and air-dry in the funnel. Weigh and determine the per cent yield and the melting point of the 1-naphthol (lit. mp 96°).

QUESTIONS

*1. Write out complete equations for the four alternate preparations of benzyne given at the beginning of the chapter. Show intermediates and byproducts of each reaction.

*2. In the absence of added nucleophiles or dienes, benzyne forms a dimer and trimer. Give the names and structures of these compounds.

*3. The 1,4-dihydronaphthalene-1,4-endoxide prepared in this experiment reacts with 2,3-dimethylbutadiene, followed by acid and then an oxidizing agent, to give 2,3-dimethylanthracene. Write equations showing how this product is formed.

*4. Write, equations involving benzyne intermediates for the following reactions:

    a. o-bromoanisole $\xrightarrow[\text{liq } NH_3]{\text{KNH}_2}$ m-anisidine

    b. o-fluorobromobenzene + Mg + anthracene → triptycene

    c. 3-(m-chlorophenyl)propionitrile $\xrightarrow{\text{KNH}_2}$ 1-cyanobenzocyclobutene

5. Suggest why six water rinses of the ether layer are required in the isolation of the 1,4-dihydronaphthalene-1,4-endoxide.

Reference

L. F. Fieser and M. J. Haddadin, *Can. J. Chem.*, **43**, 1599 (1965).

CHAPTER 28

2,4-DINITROPHENYLHYDRAZINE

Aromatic substitution can occur by three major types of mechanisms: electrophilic, nucleophilic, and free radical. Nitration is one of the most familiar electrophilic substitutions and is usually carried out with mixed nitric and sulfuric acids. The extent of nitration is controlled mainly by two reaction variables, the concentration of water and the temperature.

Nucleophilic aromatic substitution is a useful reaction when the ring is sufficiently activated for this type of attack; two nitro groups provide a very satisfactory degree of activation, and nucleophilic substitution in 2,4-dinitrohalobenzenes is a practical preparative reaction.

This experiment illustrates two types of aromatic substitution — electrophilic and nucleophilic. Both reactions are rapid and can be completed in one laboratory period. In the nitration, some byproducts may be obtained; these are to be observed by TLC. This preparation of the hydrazine is the one used commercially and, as will be seen, is extremely easy to carry out. Hydrazine is an exceptionally reactive nucleophile compared to simple amines; the reaction of 2,4-dinitrochlorobenzene with ammonia requires prolonged heating.

EXPERIMENTS

A. 1-Chloro-2,4-dinitrobenzene. In a 125 ml Erlenmeyer flask place 5 ml of concentrated nitric acid and then add 15 ml of concentrated sulfuric acid. Swirl the flask to mix the acids and add 3.0 g of chlorobenzene. Heat the reaction mixture on the steam bath with occasional swirling for 20 minutes. During the first part of the reaction, nitric oxide fumes should be swept into the aspirator when necessary by putting the end of the aspirator tube into the neck of the flask. Pour the reaction mixture onto 50 g of ice (250 ml beaker about 1/3 full), and scratch to crystallize. (*CAUTION*: Avoid contact of this product with the skin—a burning sensation will result.)

Collect the crystals on a Büchner funnel and wash with water, then transfer the moist and slightly oily solid to a 50 ml Erlenmeyer flask. Recrystallize from methanol; use the minimum volume necessary to obtain a homogeneous solution at room temperature and chill in ice bath. Collect the solid, air-dry, and weigh. Record the yield and melting point.

Dilute the methanolic mother liquor with two volumes of water and chill. Remove a sample of the oily solid (or oil) that separates and compare this and the recrystallized material on TLC (carbon tetrachloride or chloroform, depending on activity of plates). Record the results, and if time permits, check later for the identity of the impurities in the mother liquor by running TLC with authentic samples of possible compounds.

B. 2,4-Dinitrophenylhydrazine. Dissolve the recrystallized chlorodinitrobenzene in methanol (10 ml per gram) in a 50 ml Erlenmeyer flask. In a small graduated cylinder, measure 1 ml of 85 per cent hydrazine hydrate ($N_2H_4 \cdot H_2O$) per gram of dinitro compound. (Calculate the molar proportions of N_2H_4 and dinitro compound and report the results.) Dilute the hydrazine with 5 ml of methanol and add this solution to the chlorodinitrobenzene solution. After about 10 minutes, collect the product, wash with a little methanol, air-dry, and record the yield of 2,4-dinitrophenylhydrazine.

Place a small portion (about 100 mg) of the product in an 18×150 mm test tube, add 4 ml of water, and stir to moisten and suspend the solid. Add concentrated HCl dropwise, until a definite color change has occurred, then dissolve the solid by heating over a burner; add a little more water if necessary. Cool and collect crystals in a Hirsch funnel (do not wash). Record observations and identify this material. Add a small sample of this new solid to one ml of water and note any change in appearance. Add a few drops of concentrated HCl to redissolve the solid and then add a few drops of acetone. Record your observations.

Save the remaining 2,4-dinitrophenylhydrazine for later use. A convenient reagent for qualitative tests for carbonyl compounds can be prepared by dissolving 1 g of 2,4-DNPH in 5 ml of concentrated sulfuric acid and adding this to a mixture of 8 ml of water and 25 ml of ethanol.

QUESTIONS

⋆1. Write an equation showing the role of sulfuric acid in the nitration reaction.

⋆2. What other products do you expect to find in the mother liquor from the chlorodinitrobenzene?

3. The reaction of a polypeptide with 2,4-dinitrofluoro-benzene is used in an analytical procedure for identifying the terminal amino acid in the chain. After reaction, the amide bonds are hydrolyzed, liberating all but the N-terminal amino acid (containing R_1).

$$NH_2\!-\!\underset{R_1}{CH}\!-\!\underset{O}{C}\!-\!NH\!-\!\underset{R_2}{CH}\!-\!\underset{O}{C}\!-\!NH\!-\!\underset{R_3}{CH}\!-\!CO_2H\ +\ \text{(2,4-dinitrofluorobenzene)}\ \longrightarrow\ A$$

$$A\ \xrightarrow{\ H_3O^+\ }\ B + NH_2\!-\!\underset{R_2}{CH}\!-\!CO_2H + NH_2\!-\!\underset{R_3}{CH}\!-\!CO_2H$$

Write structures for compounds A and B. Suggest how this reaction sequence permits the identification of the amino acid containing R_1.

4. Write equations for all changes observed in the sequence: red 2,4-DNPH $\xrightarrow{\text{HCl}}$ yellow $\xrightarrow{\text{H}_2\text{O}}$ red $\xrightarrow{\text{HCl, acetone}}$ final solid. What do these observations teach regarding the basicity of 2,4-dinitrophenylamines?

29
ALDOL CONDENSATION

The aldol condensation is a reversible reaction in which the equilibrium normally lies to the left; special techniques, such as removal of the product from contact with the basic catalyst, are therefore needed to prepare the β-hydroxycarbonyl product. A subsequent reaction which can accompany the aldol condensation is β-elimination of the hydroxyl group, leading to the α,β-unsaturated carbonyl product. This reaction is also reversible, but in many cases, particularly with R_3 or R_4 aryl, the unsaturated product can be obtained in good yield.

$$R_2\overset{\overset{\displaystyle O}{\|}}{C}CH_2R_1 \quad \underset{}{\overset{OH^-}{\rightleftarrows}} \quad R_2\overset{\overset{\displaystyle O}{\|}}{C}\overset{-}{C}HR_1 \quad \overset{R_3\overset{\overset{\displaystyle O}{\|}}{C}R_4}{\rightleftarrows} \quad R_2\overset{\overset{\displaystyle O}{\|}}{C}-CH-\underset{\underset{\displaystyle R_4}{|}}{\overset{\overset{\displaystyle R_3}{|}}{C}}-OH$$

$$\underset{R_1}{|}$$

$$OH^- \Updownarrow \qquad \text{aldol}$$

$$R_2\overset{\overset{\displaystyle O}{\|}}{C}-\underset{\underset{\displaystyle R_1}{|}}{C}=\underset{\underset{\displaystyle R_4}{|}}{\overset{\overset{\displaystyle R_3}{|}}{C}} \quad \rightleftarrows \quad R_2\overset{\overset{\displaystyle O}{\|}}{C}-\underset{\underset{\displaystyle R_1}{|}}{\overset{-}{C}}-\underset{\underset{\displaystyle R_4}{|}}{\overset{\overset{\displaystyle R_3}{|}}{C}}-OH$$

The formation of the dehydrated aldol product involves several steps, and the detailed mechanism may vary, depending on which step is rate-determining in a given case. In the reaction of a ketone and an aromatic aldehyde, the slow step in aldol formation is the attack of the enolate at the aldehyde carbonyl group. The relative rates of aldol formation and the subsequent steps leading to the unsaturated ketone depend on several factors, and the overall process from ketone to final product presents a complex kinetic picture.

It is possible to obtain a qualitative indication of relative reactivities in the overall sequence by comparison of one aldehyde in the reaction with several ketones and *vice versa*, using standardized conditions. Provided that the products have comparable solubilities, the time required for separation of the unsaturated ketones from solution serves as a rough indication of the reactivity.

EXPERIMENTS

In this experiment, the base-catalyzed condensation of several ketones with benzaldehyde is to be compared and conclusions drawn about the extent of the reaction and the relative reactivity of the ketones.

> **Procedure.** In each of five test tubes labeled 1 to 5 place 5 ml of benzaldehyde-NaOH solution (side shelf) [0.5N in benzaldehyde and 0.25N in NaOH in 1:1 water-ethanol]. To the first four tubes add in turn 0.1 ml (5 drops) of the following ketones, noting the time that each addition is made and the time (0.2–15 min) at which a turbidity or precipitate appears: (1) acetone, (2) acetophenone, (3) cyclohexanone, (4) cyclopentanone. To the fifth tube, add *1 ml* of acetone. A reactivity order is to be established from the first four reactions; if a point is "missed" or uncertain, repeat the questionable reaction.
>
> Collect the product from tube 1, wash the crystals on the funnel with water, and recrystallize the material from aqueous ethanol (dissolve in a few ml of hot ethanol and add a few drops of water to give turbidity). Dry the crystals and determine the melting point.
>
> To tube 5 add 5 ml of water and 2 to 3 ml of ether, shake, and allow the layers to separate. Spot a TLC plate with this ether solution and also a solution of the crystals from tube 1. Develop the plate with ether-hexane (1:1).
>
> Separate the remaining ether layer, evaporate it in a test tube, convert the oil to the 2,4-dinitrophenylhydrazone (see p. 174), recrystallize from ethanol, and determine the melting point. If time permits, the product from tube 1 can also be converted to the dinitrophenylhydrazone.
>
> Deduce probable structures of the products formed in tubes 1 and 5, and confirm by consulting handbook tables for the melting points of the products and derivatives. Account for the difference in the course of the two reactions.
>
> The reaction products from cyclopentanone and cyclohexanone (tubes 3 and 4) are analogous to that formed in the reaction with acetone in tube 1; if time permits, confirm this conclusion by isolation of one or both of these products and comparison of the melting points with the literature values.

QUESTIONS

*1. Write equations for the products that would be formed in the following reactions:
   a. acetophenone plus base
   b. formaldehyde plus isobutyraldehyde plus base
   c. 1,3-diphenyl-2-propanone + benzil $(C_6H_5COCOC_6H_5)$ + base
   (The product is a highly reactive diene in Diels-Alder reactions.)

*2. Aldol condensations and dehydrations can also be effected by acid catalysis. Give structures and mechanisms of formation of the following products:
   a. $(CH_3)_2C=O \xrightarrow{HCl} C_6H_{10}O + C_9H_{14}O$
   b. $(CH_3)_2C=O \xrightarrow{H_2SO_4} C_9H_{12}$

*3. Given that the rate of aldol condensation of aceto-phenone and benzaldehyde depends on the concentrations of both organic reactants and the base, suggest which of the steps in the overall reaction can be rate-determining.

4. In several instances, the relative stability of exocyclic and endocyclic isomers in cyclic systems is that indicated by the equilibria A and B:

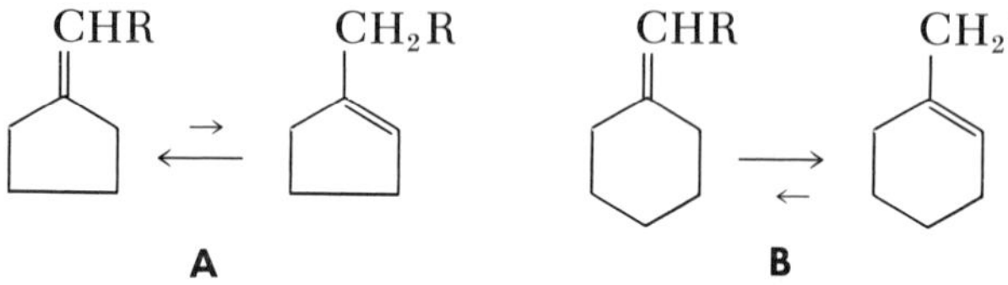

A   B

Suggest how these relationships might account for the relative rates of condensation observed with cyclohexanone and cyclopentanone, assuming that the rate determining step is:
   a. the condensation of anion and aldehyde.
   b. the dehydration of the aldol.

5. Suggest the order of reactivity of the following alde-hydes in the base-catalyzed aldol condensation with aceto-phenone; explain the reason for your sequence.

Benzaldehyde	p-Anisaldehyde
	(p-methoxybenzaldehyde)
p-Tolualdehyde	p-Chlorobenzaldehyde
p-Nitrobenzaldehyde	p-Dimethylaminobenzaldehyde

CHAPTER 30

PREPARATION AND SEPARATION OF o- AND p-NITROPHENOL

Electrophilic substitution in the highly activated phenol ring occurs under very mild conditions. For mononitration, dilute aqueous nitric acid is used, although this reagent functions only as an oxidizing agent for aromatic hydrocarbons. Treatment of phenol with the typical nitric acid-sulfuric acid mixture gives only di- and trinitro derivatives.

The separation of o- and p-nitrophenol depends on the volatility of the *ortho* isomer due to intramolecular hydrogen bonding (chelation). In the *para* isomer, intermolecular hydrogen bonding leads to association of molecules in the liquid and much lower vapor pressure. On steam distillation of the mixture, the *ortho* isomer is obtained in very pure form in the distillate. The *para* isomer can then be isolated from the nonvolatile residue, which contains polynitro compounds and oxidation products.

179

EXPERIMENT

In a 250 ml Erlenmeyer flask place 10 ml of concentrated nitric acid and 35 ml of water. Weigh out 8.0 g of phenol ("loose crystals") in a 50 ml beaker. (*CAUTION:* Avoid contact with skin.) Add 2 ml of water and allow the mixture to liquefy. With a disposable pipette, add 1- to 2-ml portions of the phenol to the nitric acid and cool as necessary by swirling in a pan of cold water to keep the temperature of the reaction mixture at 45 to 50°. After all of the phenol has been added (rinse beaker with 1 ml of water), shake the flask intermittently for 5 to 10 minutes while the contents cool to room temperature. Meanwhile, assemble apparatus for steam distillation.

Transfer the reaction mixture to a separatory funnel and drain the oily organic layer into a 3-neck 500 ml flask. Add 150 ml of water and then carry out steam distillation (see Fig. 8.1) until no further o-nitrophenol appears in the distillate. Collect the o-nitrophenol, air dry, and determine yield and melting point.

For isolation of the *para* isomer, adjust the total volume of the distillation residue to about 200 ml by adding more water or removing water by distillation (burner). Decant the hot mixture through a coarse fluted filter or loose cotton pad, add about 1 to 2 g of charcoal to the hot filtrate, heat again to boiling, and refilter to remove charcoal. Cool a 500 ml Erlenmeyer flask in ice and pour into it a small portion of the hot solution to promote rapid crystallization and prevent separation of the product as an oil. Then add the remainder of the solution in small portions so that each is quickly chilled. Collect crystals, dry, and report yield and melting point.

QUESTIONS

1. Why is steam distillation, as opposed to simple distillation of the oily product mixture, preferred for this separation?

2. The solubility of phenol in water is 9 g/100 ml at room temperature; explain how 8 g of phenol and 2 ml of water can form a homogeneous solution.

3. Account for the differences in the following properties of o- and p-nitrophenol.

	ORTHO	PARA
Solubility in H_2O (g/100)	0.20	1.70
Infrared ν_{OH} ($CHCl_3$), cm^{-1}	3200	3530

CHAPTER 31

SYNTHESIS OF LIDOCAINE

Local anesthetics (pain killers) are an important and well studied class of synthetic drugs. Some common local anesthetics are shown below. Of these, only cocaine* is a naturally occurring compound, and the synthetic drugs are used to avoid the addictive narcotic effects of the former.

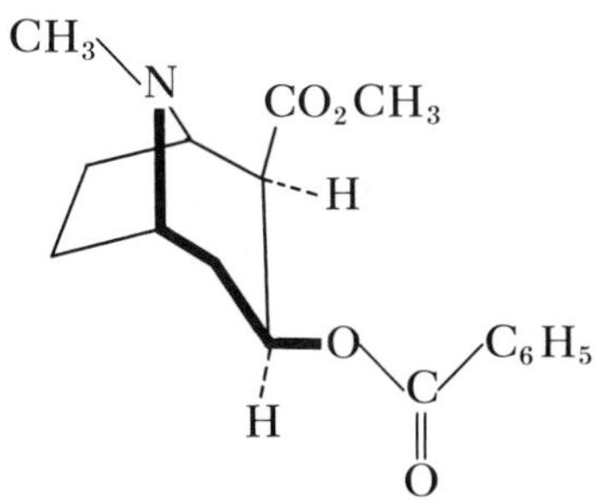

Cocaine

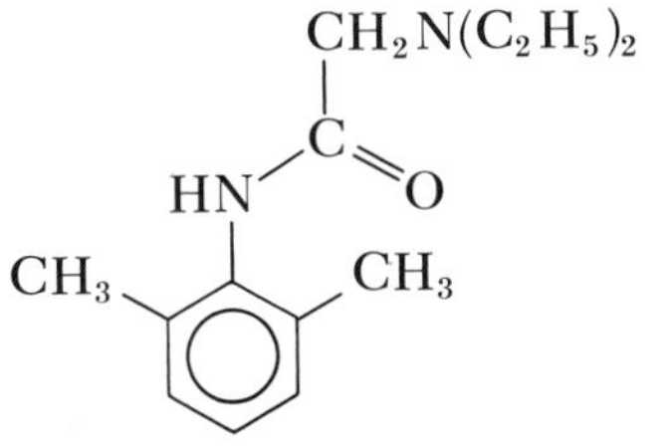

Lidocaine
(Xylocaine)

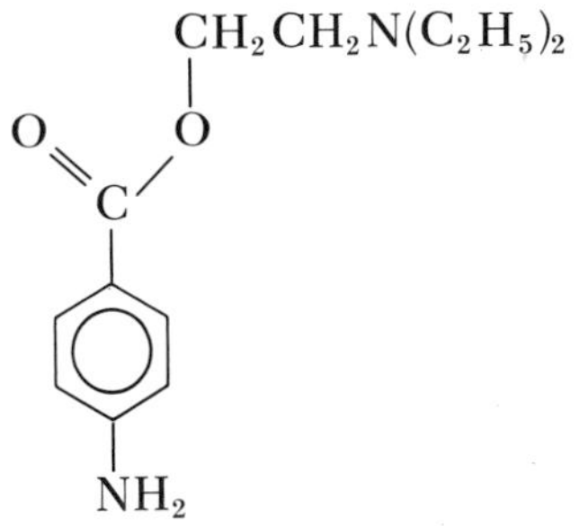

Procaine
(Novocaine)

In most of the hundreds of local anesthetics that have been synthesized, two structural features are prominent: the compounds are benzoate esters or anilides and contain a dialkylamino group separated by 1–4 atoms from the carbonyl center, as indicated in the structure of cocaine. The dialkylamino group is a characteristic unit in the structures of many diverse medicinal agents such as antihistamines, antimalarial compounds, and tranquilizers.

In this experiment the local anesthetic lidocaine will be synthesized and isolated in the form of its hydrochloride. Lidocaine hydrochloride (the generic name) is sold under various trade names, the most common of which is Xylocaine. It is noted for its relatively high anesthetic activity

---

*The suffix *caine* used so commonly with synthetic local anesthetics is derived from cocaine, which in turn is a combination of *coca-* + *-ine*, meaning a nitrogenous compound from coca plants. The original pronunciation of cocaine (kō′kȧ-ēn) has evolved into kō·kān.

when applied to the skin or injected into nerves and it has a low toxicity and incidence of side effects.

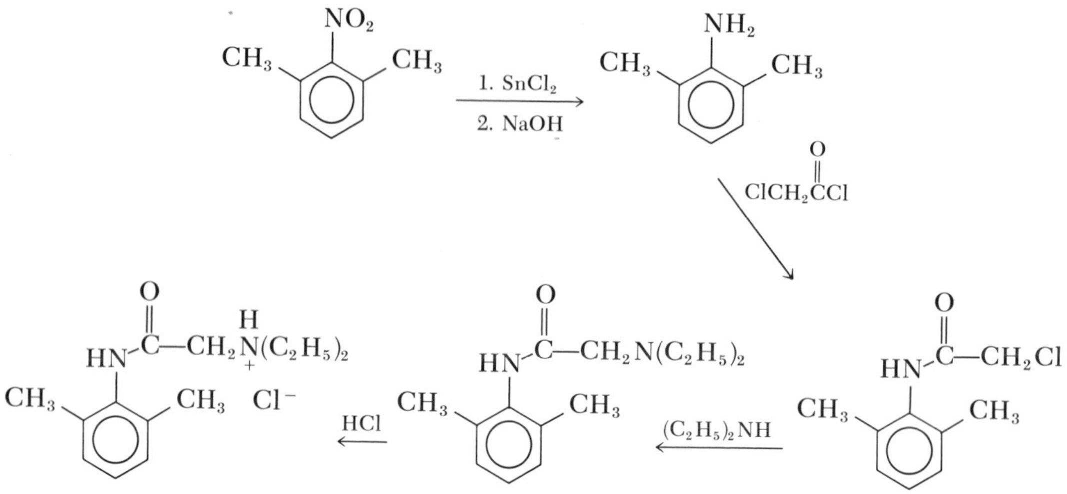

mp 127–129°
Monohydrate
mp 77–78°

mp 68–69°

mp 145–146°

This synthetic sequence illustrates several important reactions. The reduction of an aromatic nitro compound is most commonly accomplished with metals such as iron, zinc, or tin. Stannous chloride is more rapid and convenient because the reaction is homogeneous. The second and third steps in the synthesis illustrate the very large difference in reactivity of the two electrophilic centers in chloroacetyl chloride.

EXPERIMENTAL PROCEDURE

2,6-Dimethylaniline (2,6-xylidine)

Dissolve 5.0 g of 2,6-dimethylnitrobenzene in 50 ml of glacial acetic acid in a 250 ml Erlenmeyer flask. In a 125 ml flask dissolve 23 g of $SnCl_2 \cdot 2H_2O$ in 40 ml of concentrated hydrochloric acid, heating on a steam bath if necessary. Add the $SnCl_2$ solution in one portion to the nitroxylene solution, swirl to mix, and let the mixture stand for 15 minutes. Cool the mixture and collect the crystalline salt in a Büchner funnel. Transfer the moist crystals to an Erlenmeyer flask, add 25 ml of water, and make strongly basic by adding 30% KOH solution (40–50 ml required). After cooling, extract with 25 and 10 ml portions of ether, rinse the ether extracts twice with 10 ml of water, and dry over K_2CO_3. Evaporate the dried and filtered solutions to an oil, transfer and rinse into a tared 25 × 100 mm test tube, complete evaporation, weigh, and calculate the per cent yield of 2,6-dimethylaniline.

α-Chloro-2,6-dimethylacetanilide (α-chloroaceto-2,6-xylidide)

To a 250 ml Erlenmeyer flask add the xylidine, 25 ml of glacial acetic acid, and 3.7 g (2.6 ml) of chloroacetyl chloride, in that order. Warm the solution on a steam bath to 40–50°, remove, and add a solution of 5 g of sodium acetate in 100 ml of water. Cool the mixture and collect the product in a Büchner funnel. Rinse the solid in the funnel with water until the acetic acid odor is gone, and dry as much as possible by pressing and drawing air through the filter cake in the funnel. Transfer the product to a disk of filter paper and let it air-dry until the next laboratory period. Weigh, determine the melting point, and calculate the per cent yield.

α-Diethylamino-2,6-dimethylacetanilide

In a 100 ml round-bottom flask place the chloroaceto-xylidide obtained from the preceding experiment and 45 ml of toluene; then add three moles of diethylamine (d 0.71) per mole of xylidide. (Save out a few mg of the starting material for TLC comparison). Fit the flask with a reflux condenser, add a boiling stone, and reflux vigorously over a small flame. The progress of the reaction can be measured conveniently at 15 to 30 minute intervals by TLC. (Turn off the flame and, after boiling stops, remove a sample with a fine capillary; add a fresh boiling stone and resume the refluxing. Spot the solution and the starting material on a silica gel plate and develop with chloroform).

After the starting material is gone (TLC) or after 90 minutes refluxing, whichever comes first, cool the mixture and filter out the crystals; rinse them with a small amount of pentane, air-dry, and weigh. (see questions 4 and 5).

Transfer the filtrate to a separatory funnel, and extract with two 25-ml portions of 3N HCl. Cool the acidic aqueous layer in an Erlenmeyer flask and add 30 per cent aqueous KOH until the solution is strongly basic. Extract with 20 ml of pentane. Rinse the pentane layer with six 10-ml portions of water, dry over K_2CO_3, and concentrate in a 125 ml Erlenmeyer flask to an oil. Crystallize the product by cooling in an ice-salt bath and scratching. The lidocaine is very soluble in pentane and will redissolve on warming. It can be isolated at this point by rapid filtration, or it can be converted to the hydrochloride.

To convert the base to the hydrochloride, dissolve the lidocaine in 20 ml of anhydrous ether. In the hood, bubble HCl gas from a tank or H_2SO_4/NaCl generator* through a 6-inch length of glass tubing into the solution. Stir the mixture during the addition with the tubing and, by scratching the sides of the flask, induce crystallization of the gummy product. When the solution begins to clear, stop the HCl addition and continue scratching and rubbing the product to break up the lumps. After a granular solid is obtained, collect the product in a Hirsch funnel and rinse with several 1- to 2-ml portions of ether to remove excess HCl. Air-dry, determine the melting point, weigh, and calculate the per cent yield of lidocaine hydrochloride.

---

*This is conveniently assembled from a 1000 ml suction flask containing granular NaCl and fitted with a 1-hole stopper with a separatory funnel containing concentrated H_2SO_4. A length of rubber tubing is attached to the side arm, and the rate of HCl evolution is controlled by the rate of H_2SO_4 addition, and occasional mixing by shaking.

QUESTIONS

⋆1. Elemental analysis of the crystalline salt isolated from the $SnCl_2$ reduction gave the following results: 33.4 %C, 4.2 %H, 36.8 %Cl, 4.9 %N, 20.6 %Sn. Calculate its empirical formula and suggest a structure.

⋆2. Write a balanced equation for the $SnCl_2$ reduction. Compare your answer with the actual molar ratio of reactants used.

⋆3. What is the function of the sodium acetate in the second step of the synthesis? Write a complete equation for the reaction.

4. What is the compound which crystallized from the refluxing toluene in the last step? Write a balanced equation for the reaction.

5. Compare the per cent yields of lidocaine and the compound in question 4. Which is likely to be the more accurate measure of lidocaine formed in the reaction and why?

6. Because of its importance as a local anesthetic, there are many patents describing syntheses of lidocaine. Using *Chemical Abstracts*, find one such patent in which the synthetic method is significantly different from the one you used.

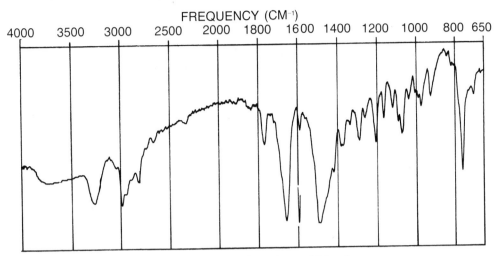

FREQUENCY (CM⁻¹)

FIGURE 31.1 Infrared spectrum of lidocaine.

7. The infrared spectrum of lidocaine is reproduced in Figure 31.1. Assign as many bands as possible to functional groups and other structural units in the molecule.

Reference

E. J. Ariens, A. M. Simonis, and J. M. van Rossum, Chemical and Physical Properties of Drugs with Local Anesthetic Action. *In* E. J. Ariens (Ed.), *Molecular Pharmacology*, Vol. I. Academic Press, New York, 1964, pp. 352–363.

32

SYNTHESIS OF
SULFANILAMIDE

Sulfa drugs, like local anesthetics (Chapt. 31), comprise a group of compounds having a key structural feature which imparts a specific pharmacological property. In the sulfa drugs, this unit is the *p*-amino-benzenesulfonamide group. The parent compound and a number of derivatives such as sulfathiazole and sulfapyridine, with a heterocyclic ring substituted on the amide nitrogen, are antibacterial agents useful against streptococcal and pneumonococcal infections.

Sulfanilamide   Sulfapyridine   Sulfathiazole

The compounds act by competitively inhibiting the incorporation of *p*-aminobenzoic acid. This acid is essential for growth of the micro-organisms, and the structurally similar sulfonamides, with a pK_a similar to that of the carboxylic acid, block the metabolic pathway. Thousands of analogs of sulfanilamide have been prepared and tested for antibacterial properties. A number of these compounds are used clinically, although they have been replaced to a considerable extent by naturally occurring antibiotics, such as penicillin and streptomycin.

The preparation of the parent compound, sulfanilamide, is a simple multistep process, involving introduction and removal of an acetyl blocking group to control the course of the synthesis (see Chapt. 7). The three steps, starting with acetanilide (see Chapt. 2), can be carried out without purification of intermediates and can be completed in a single 3 hour laboratory period.

EXPERIMENTAL PROCEDURE

p-Acetamidobenzenesulfonyl Chloride

In a 125 ml Erlenmeyer flask place 5.0 g of acetanilide. To permit addition of the chlorosulfonic acid in one portion, the surface area of the acetanilide is reduced by melting it in the flask over a soft flame and then rotating the flask to spread the material in a thin film over the bottom and lower 1/4 inch of the walls. This step may be done in advance; stopper the flask securely and store.

Prepare a gas trap for HCl fumes with tubing from flask to a funnel and invert over a beaker of water as used in the Friedel-Crafts reaction (page 119). In a *dry* graduate or test tube (depending on the dispensing arrangements), obtain 13 ml of chlorosulfonic acid, taking exactly the required amount. (*CAUTION*: This reagent is extremely hazardous and will cause severe burns. DO NOT pour any quantity of the liquid into a sink.) Cool the acetanilide in an ice bath, add the chloro-sulfonic acid in one portion, and connect the flask to the trap. The flask should be removed from the ice bath and returned only if the reaction becomes too vigorous. After 10 to 15 minutes, when most of the solid has dissolved, warm the flask on the steam bath; continue heating for about 10 minutes after the solid has disappeared. Cool the liquid (trap still connected) and then pour onto 75 to 100 ml of ice. (Add water to the flask and transfer the residue to the main batch.) Stir and rub the solid sulfonyl chloride with a spatula until the lumps are broken up, collect on Büchner funnel, and wash with a little water.

p-Acetamidobenzenesulfonamide

Transfer the moist sulfonyl chloride to the original 125 ml Erlenmeyer flask. Add 40 ml of concentrated aqueous ammonia and stir and shake. As the reaction proceeds, the granular sulfonyl chloride gives way to a thin paste of the amide. Warm for 5 minutes on the steam bath, then cool and collect the solid. Wash with water, press, and drain thoroughly.

Sulfanilamide

Transfer the cake of acetamidosulfonamide to the same Erlenmeyer flask and add 10 ml of $6N$ hydrochloric acid (5 ml concentrated HCl plus 5 ml water). Heat to boiling over a low flame until all solid has dissolved and continue gentle boiling for 10 minutes. Cool the solution and add water to replace that lost in evaporation; add activated carbon and filter through paper. The filtrate is then treated with solid sodium bicarbonate, with stirring, until the pH is neutral. Collect the solid sulfanilamide, wash with water, and air-dry. Label with melting point and yield and submit the product to your instructor.

QUESTIONS

1. What product would be expected if aniline rather than acetanilide were treated with chlorosulfonic acid?

2. What is the compound in solution after boiling the acetamidosulfonamide in hydrochloric acid? What would be the result if excess sodium hydroxide solution were used for the final neutralization after acid hydrolysis?

3. Write a balanced equation for the overall reaction of acetanilide and chlorosulfonic acid and indicate the mechanism of the reaction(s).

CHAPTER 33

NITROSATION AND DIAZOTIZATION

One of the most important aspects of the chemistry of amines is their reaction with nitrous acid. This unstable reagent is generated by addition of sodium nitrite to mineral acid at 0°; on standing or brief warming, the acid decomposes to oxides of nitrogen.

$$2 \text{ HONO} \xrightarrow{-H_2O} N_2O_3 \longrightarrow NO + NO_2$$

Reactions of nitrous acid are carried out in the presence of strong acid to generate the reactive electrophilic species, which can be represented as the nitrosonium ion NO^+. The reaction with amines is rapid even in strong acid in which the amine is present largely as the unreactive ammonium ion; attack of NO^+ occurs with the small equilibrium concentration of free base.

With any amine, NO^+ is coordinated, like a proton, by the unshared electron pair. With a primary or secondary amine, a proton can be lost from nitrogen to give an N-nitroso compound R_2N—NO. This is the end product in the latter case, but with a primary amine, tautomerization of the nitroso compound RNHNO occurs rapidly and irreversibly to give a diazo hydroxide which in acid solution undergoes protonation and loss of water to the diazonium ion.

$$\underset{\underset{|}{H}}{-\overset{H}{\underset{|}{N}}} : + NO^+ \longrightarrow -\overset{H}{\underset{|}{N}}{}^+ - N{=}O \xrightarrow{H_2O} -\ddot{N} - N{=}O + H_3O^+$$

$$\underset{\underset{|}{H}}{R - N - N{=}O} \longrightarrow R - N{=}N - OH \xrightarrow{H^+} R - \overset{+}{N}{\equiv}N + H_2O$$

Tertiary aromatic amines undergo nitrosation at a free *ortho* or *para* **189** position to give the C-nitroso compound:

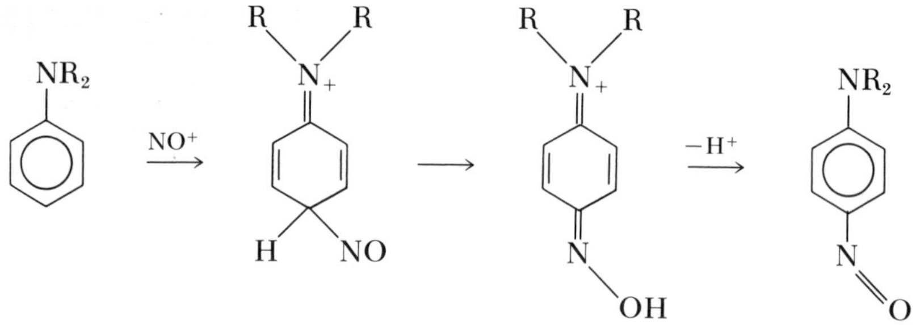

DIAZONIUM COMPOUNDS

Aryl diazonium ions are highly versatile reagents with many synthetic uses. The two main types of reaction are displacement by a variety of nucleophiles, with loss of nitrogen, and coupling. In the latter reactions the diazonium ion behaves as an electrophile similar to NO^+.

Displacement: $Ar\overset{+}{N}{\equiv}N + X^- \longrightarrow ArX + N_2$

Coupling: $Ar\overset{+}{N}{\equiv}N + Z^- \longrightarrow ArN{=}N{-}Z$

In the preparation of phenol by displacement, the diazonium salt is prepared using the weakly nucleophilic sulfate ion, so that the aryl carbonium ion combines with water.

$$Ar\overset{+}{N}{\equiv}N + HSO_4^- + H_2O \longrightarrow ArOH + N_2 + H_2SO_4$$

One of the important coupling reactions occurs with activated aromatic rings such as phenols. A sensitive qualitative test for the diazonium ion is the coupling with 2-naphthol to form an azo dye. With an amine, coupling leads to the diazoamino compound, or triazene.

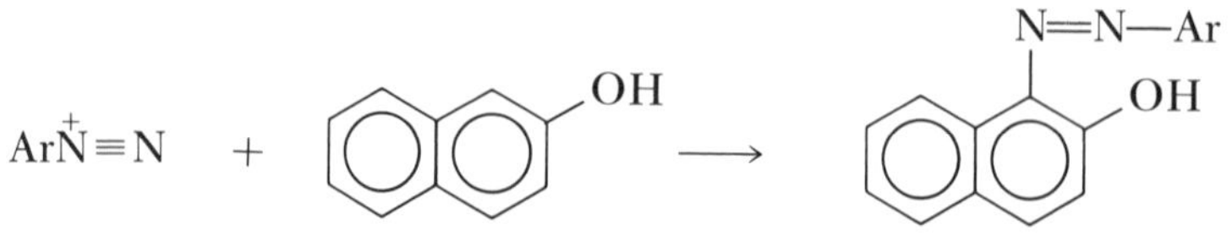

$$Ar\overset{+}{N}{\equiv}N + RNH_2 \longrightarrow ArN{=}N{-}NHR$$

triazene

EXPERIMENTAL PROCEDURES

The behavior of several types of amines with nitrous acid will be compared, and two small-scale preparative reactions with benzene-diazonium ion will be carried out.

Diazotization of Amines

In each of four test tubes place 1 ml of $6N$ hydrochloric acid and add the following: (1) 0.4 ml of aniline, (2) 0.2 ml of aniline, (3) 0.2 ml of n-butylamine, (4) 0.2 ml of N-methylaniline. Place the labeled tubes in an ice bath and allow to cool to 0°, and then treat each in turn dropwise with 1.0 ml of $2N$ $NaNO_2$ solution. Record the observed behavior and write equations.

Add a few drops of solution from tubes (2) and (3) to 1-ml portions of alkaline 2-naphthol solution, record results, and write an equation for any reaction.

To tube (1) add about 2 g of sodium acetate dissolved in a little water. Collect the crystalline precipitate on a Hirsch funnel, wash with water, and recrystallize the moist cake from the minimum volume of ethanol or methanol and record the melting point. The melting point of 1,3-diphenyltriazene is 96–98°. Account for the formation of this product and write equations.

Preparation of Phenol

In a 100 ml round-bottom flask place 20 ml of $4N$ H_2SO_4 and add 2.0 ml of aniline. The resulting suspension is cooled to 0–5° in ice, and then 10 ml of $2N$ $NaNO_2$ solution is added dropwise or in small portions, with swirling. After the addition of the nitrite, check for the presence of excess HNO_2 with iodide-starch paper; an immediate blue-black color indicates that excess nitrous acid is present.

$$2HNO_2 + 2HI \longrightarrow 2NO + I_2 + 2H_2O$$

If the KI-starch reaction is not positive, add an additional 1 ml of $NaNO_2$ solution.

Remove the solution from the ice bath and allow it to stand for 5 to 10 minutes; meanwhile assemble a simple distillation set-up. Clamp the flask above a wire gauze and, before attaching the distillation head, place a thermometer in the solution and warm with a flame to 45°. Remove the flame and watch the temperature. After 10 minutes, or sooner, if the temperature drops, resume heating and distill. Collect about 25 ml of distillate in a 50 ml Erlenmeyer flask; an adapter is not necessary. Add about 10 to 15 g (about a tablespoon) of salt to the distillate. The volume of the oily upper layer can be estimated by decanting into a 10 ml graduated cylinder. Test the solubility of the oily layer in dilute HCl, dilute NaOH and $NaHCO_3$ solution; test the reaction of a few drops of the oil with 1-ml portions of aqueous ferric chloride and Br_2—CCl_4 solution. Write equations for all of the reactions observed.

p-Nitroso-N,N-dimethylaniline

In a 25 × 100 mm test tube dissolve 1 ml of N,N-dimethylaniline in 7 ml of 6N HCl and cool to 0° in an ice bath. Chill 4 ml of 2N $NaNO_2$ solution and add this dropwise to the cold amine solution. Collect the resulting solid on a Hirsch funnel and drain thoroughly but do not wash. Return the solid to the test tube, dissolve in a minimum volume of water and add 1N NaOH dropwise until the color change is complete. Extract the mixture with a small volume of ether, dry the ether solution, evaporate, and collect the solid. Write equations with structures of both solids and account for the colors of the compounds.

QUESTIONS

⋆1. Write Lewis electron structures for the following molecules or ions:
  a. N_2O_3
  b. NO
  c. NO_2
  d. H_2ONO^+
  e. $C_6H_5N_2^+$
  f. NO_2^+

⋆2. The reaction of tri-n-propylamine with nitrous acid, followed by warming the acid solution, gives propionaldehyde and di-n-propylamine plus inorganic products. Suggest a mechanism by which this reaction might occur, showing the inorganic product originating from nitrous acid.

⋆3. Write three products that might be expected to arise from the reaction of n-propylamine plus nitrous acid.

4. In a misguided effort to test for the presence of a carbonyl group, a student added a sample of aniline to the ethanolic 2,4-DNPH reagent described on page 174 for qualitative identification. A voluminous precipitate was observed. What was this precipitate?

5. In the preparation of phenol from benzenediazonium ion, sulfuric acid is preferable to hydrochloric acid for several reasons. Suggest why.

6. 1,3-Diphenyltriazene on treatment with an acid catalyst (usually aniline hydrochloride) undergoes rearrangement to a red isomer which can be diazotized and coupled with 2-naphthol. Suggest a structure for this red isomer.

CHAPTER 34

HETEROCYCLIC SYNTHESES

The most important heterocyclic compounds are those containing nitrogen and other heteroatoms in a fully unsaturated five- or six-membered ring. Such ring systems are isoelectronic with cyclopentadiene anion and benzene, respectively, and are aromatic compounds; typical examples are pyrrole, pyrazole, isoxazole, pyridine, and pyrimidine. These and other rings are present in important natural substances such as hemoglobin, vitamins, and nucleic acids, and they also form the basic structure of many synthetic drugs and dyes. Rings with a number of combinations of heteroatoms are readily prepared, and heterocyclic compounds comprise the bulk of all known organic molecules.

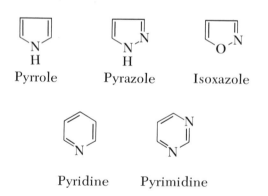

| Pyrrole | Pyrazole | Isoxazole |

Pyridine   Pyrimidine

Syntheses of many heterocyclic compounds are based on simple reactions such as condensation of carbonyl groups with amino and activated methylene groups in bifunctional molecules. The experiments in this chapter demonstrate how a number of the fundamental heterocyclic rings can be built from a single starting material, acetylacetone. The β-diketone system in acetylacetone is largely enolized, and the compound and its reactions should be represented with this tautomeric form. This enolic ketone contains two electrophilic centers and a highly reactive nucleophilic central carbon. Condensations can occur between any two of these functional groups and two complementary groups in another molecule or molecules, as indicated schematically in A and B.

$$
\begin{array}{cc}
\text{A} & \text{B}
\end{array}
$$

PART A

The first group of syntheses to be carried out are condensations of type A, in which two nucleophilic centers in the second component condense with the dicarbonyl system, eliminating two moles of water. Typical examples are the formation of the pyrazole (**1**) and the pyrimidine (**2**), illustrated with the enolic dibenzoylmethane.

(34.1)

1	Dibenzoylmethane	2
1,3,5-Triphenylpyrazole		2-Methyl-4,6-diphenylpyrimidine

Condensations of this type are very simple to carry out, and with the very reactive acetylacetone, these preparations involve little more than mixing the reagents in an appropriate medium. When the solubilities of starting materials and products permit, water is a very effective solvent. The rate of these carbonyl condensations in aqueous solution is pH dependent; a slightly basic solution is optimum for the reactions studied here.

Procedures are indicated in the following paragraphs for reactions of acetylacetone with the four reagents listed alphabetically in Table 34.1. The names of the products are given, also in alphabetical order, in Table 34.2. Before proceeding with the experiment, determine the product that will be obtained from each reagent (see Question 2 at end of Chapter). Then carry out one or more of the reactions as directed by your instructor. Only minimum directions are given; use your own judgment and experience in isolating the products. Submit products, labeled with yield, melting point and structural formula, to your instructor.

TABLE 34.1

STARTING MATERIALS	FORMULA
Cyanoacetamide	$NCCH_2CONH_2$
Guanidine	$(NH_2)_2C{=}NH$
Hydrazine	NH_2NH_2
Hydroxylamine	NH_2OH

TABLE 34.2

PRODUCTS

2-Amino-4,6-dimethylpyrimidine, mp 156° (from MeOH)
3-Cyano-4,6-dimethyl-2-pyridone, mp 290° (from H_2O)
3,5-Dimethylisoxazole, bp 142°
3,5-Dimethylpyrazole, mp 107° (from H_2O)

PROCEDURES

1. Dissolve 0.84 g of cyanoacetamide and 0.5 g of sodium carbonate in 15 ml of water and add 0.01 mole (1 ml) of acetylacetone.

2. Dissolve 1.8 g of guanidine carbonate and 1 g of sodium carbonate in 5 ml of water and add 2 ml of acetylacetone. After partial air-drying, recrystallize the product from a minimum volume of methanol; filter the hot solution to remove traces of inorganic solid.

3. Mix 1 ml of 85 per cent hydrazine hydrate and 5 ml of water and add 1 ml of acetylacetone.

4. Dissolve 2.8 g of hydroxylamine hydrochloride and 3.3 g of anhydrous sodium acetate in 5 ml of water and add 2 ml of acetylacetone.

PART B

Condensations of the second type (B) involve the reaction of ammonia or an amine with the enolic diketone to give an enamino ketone, —NH—C=C—C=O. These compounds are stabilized by conjugation of the —NH and —CO groups, and are better regarded as "vinylogous" amides than amino ketones. The enamine system can provide one or two nucleophilic centers for further reactions.

One important application of this type of condensation is the Knorr pyrrole synthesis. An α-amino ketone is generated in the presence of the enol, and the intermediate enamine cyclizes to a pyrrole. In the following procedure, the amino ketone is formed by reduction of isonitroso-acetophenone.

$$(34.2)$$

mp 151°

PROCEDURE

3-Acetyl-2-methyl-4-phenylpyrrole

Place 6 ml of acetic acid and 1.5 ml of water in a 50 ml Erlenmeyer flask and add 1.8 g of isonitrosoacetophenone and 1.3 ml of acetylacetone. To this solution add, in 4 portions, 2.5 g of zinc dust. Swirl vigorously between each addition, cooling in an ice bath if the flask becomes too hot to hold in the hand. After all of the zinc has been added, heat the reaction mixture to gentle boiling over a wire gauze with a low flame for 2 to 3 minutes. Decant the hot solution from residual zinc into 50 g of ice. Add alkali until $Zn(OH)_2$ just begins to precipitate and extract the mixture twice with ether. Wash the ether solution with dilute base and then water, dry over $MgSO_4$, and evaporate to about 5 ml volume. Cool and crystallize and collect the product and recrystallize from methanol. The pyrrole tends to crystallize in extremely fine needles which occlude the oily mother liquor, and thorough washing with cold solvent on the funnel (with the aspirator disconnected) is important. Determine the yield and melting point and submit the product, labeled with the compound name, structure, and melting point.

PART C

Two other condensations of this type are performed under very similar conditions, with an additional component present in one case. In these reactions, ammonia, conveniently added in the form of ammonium acetate to buffer the solution, gives the enaminoketone (Eq. 34.3). This intermediate can supply two nucleophilic centers for condensation with another molecule of acetylacetone (or a second molecule of enamine) (Eq. 34.4). If a more reactive carbonyl group is available, however, reaction of two molecules of enamine can occur at this center. In the reaction with acetaldehyde described below, the acetaldehyde can very conveniently be added in the form of its ammonia addition product, $CH_3CH(OH)NH_2$ (Eq. 34.5).

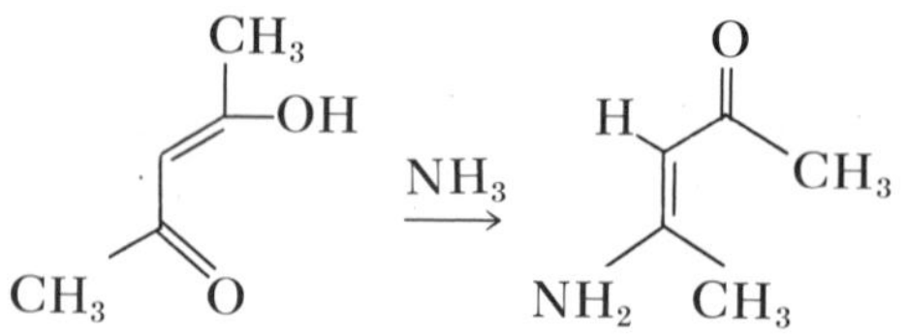

(34.3)

(34.4)

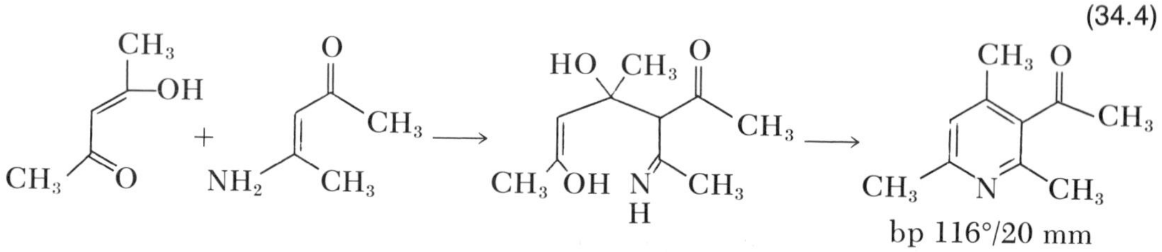

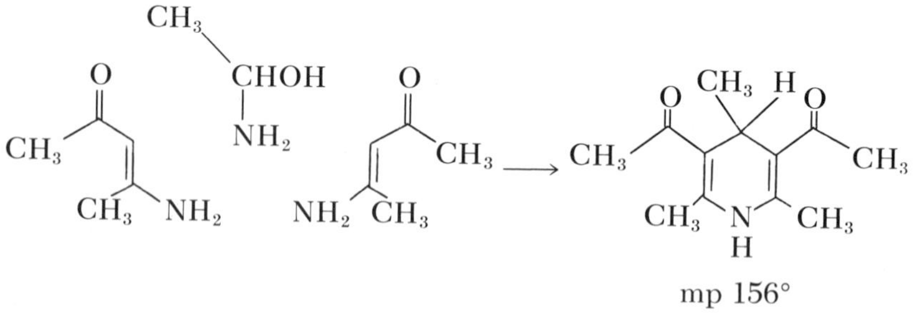

bp 116°/20 mm

(34.5)

mp 156°

PROCEDURES

3-Acetyl-2,4,6-trimethylpyridine

In a 125 ml Erlenmeyer flask mix 8 g of ammonium acetate and 10 ml of acetylacetone. Warm on the steam bath for 30 minutes, cool, and add saturated sodium carbonate solution in portions until CO_2 evolution stops. Extract the mixture with 25 ml of ether and then with 15 ml of ether. Dry the combined ether solutions (do not backwash with water) with $MgSO_4$ and evaporate to an oil. Transfer the oil with a pipette to a 50 ml round-bottom flask and set up for vacuum distillation using only a distilling head with thermometer, a vacuum take-off adapter, and a tared 50 ml round-bottom flask as a receiver; a condenser is unnecessary. Add several boiling stones, connect to aspirator suction, and distill with a low flame. Some splashing will occur, but a nearly colorless distillate can be obtained. Report the yield and boiling range of the product.

3,5-Diacetyl-2,4,6-trimethyl-1,4-dihydropyridine

Dissolve 0.6 g of acetaldehyde-ammonia and 0.8 g of ammonium acetate in 10 ml of water and add 2 ml of acetylacetone. Warm for about 2 minutes at 50° until all liquid droplets dissolve; allow to cool. Remove a few drops of the solution and chill to obtain a seed, then seed the main solution and allow to crystallize for several days at room temperature if time permits. Collect the product and report the yield and melting point. (This preparation illustrates the point of view, held by some individuals, that organic chemistry is one of the more rewarding art forms.)

PART D

A variation of the type A condensation discussed earlier is illustrated in the preparation of 2,4,6-trimethylquinoline. The intermediate enamine from an aromatic amine and the β-diketone is prepared and isolated in a first step, and this is then cyclized. The second step is a typical electrophilic aromatic substitution with the protonated carbonyl group of the enamine.

(34.6)

2,4,6-Trimethylquinoline
· $2H_2O$ (mp 62°)

PROCEDURE

4-(p-Toluidino)-3-penten-2-one

In a 25 ml distilling flask place 4.2 g of p-toluidine, 6 ml of acetylacetone and a boiling stone. Insert a thermometer in the neck so that the bulb is just immersed in the liquid. Use a 13 × 100 mm test tube as a receiver (see Fig. 35.1). Heat the flask over a wire gauze with a low flame until the temperature reaches 140° and boiling commences. Adjust the flame to maintain gentle boiling with a minimum of distillation into the side arm for five minutes, then increase the heating rate and collect water and excess acetylacetone until the temperature reaches 215°. Cool the flask and pour the contents into a 25 × 100 mm test tube and rinse with 4 to 5 ml of pentane. Chill the pentane solution in an ice bath, and after crystallization, collect the product and wash on the filter with a small volume of chilled pentane.

2,4,6-Trimethylquinoline

In a 250 ml Erlenmeyer flask place 25 ml of concentrated sulfuric acid and add 4.0 g of enaminoketone. (Reduce the amount of H_2SO_4 proportionately if less enamine is available.) Warm the solution for 20 to 25 minutes on the steam bath, then cool and add about 100 g of ice. Chill the solution in an ice bath and after crystallization of the salt, collect it on a Büchner funnel and wash with a few ml of water. Suspend the moist salt in 40 ml of water and add concentrated ammonium hydroxide in small portions (see question 6). Chill and crystallize the resulting oil, and collect the solid. Dissolve the moist solid in 10 ml of ethanol, add water until the solution is not quite turbid (about 10 ml of water), seed, stopper loosely, and allow to stand. The dihydrate of the quinoline crystallizes in long needles. Record the melting point and the yield (calculated on the basis of trimethylquinoline · $2H_2O$) and submit the product to your instructor.

QUESTIONS

*1. Explain how pyrrole qualifies as an aromatic compound according to the $(4n + 2)\pi$ electron rule. From these considerations, predict whether pyrrole or pyridine would be the more strongly basic compound.

*2. Write equations with structural formulas for reactions of the four starting materials in Table 34.1 with acetylacetone.

*3. A β-keto ester, $RCOCH_2CO_2R'$, is often used instead of a β-diketone in condensations of Type A; the product then contains a carbonyl group adjacent to the heteroatom. Write structural formulas for the products that would be obtained in the reaction of hydrazine and hydroxylamine with ethyl acetoacetate.

4. 4-Amino-3-penten-2-one, the intermediate in Equations 34.4 and 34.5, can be isolated as a low melting solid if desired. Predict the product that would be formed if this compound were heated in aqueous solution containing (a) ammonium acetate, (b) hydrazine acetate (NH_2NH_3OAc).

5. In the Knorr condensation (Eq. 34.2), no pyrrole is obtained if the acetylacetone is added after addition of the zinc. A product with the formula $C_{16}H_{14}N_2$ can be isolated instead. Suggest the structure of this product.

6. Heterocyclic bases, and amines in general, can form two types of salts with sulfuric acid: sulfate $(R_3NH)_2^+SO_4^=$ and bisulfate $(R_3NH)^+HSO_4^-$. Suggest the formula of the salt that precipitates in the quinoline preparation. Describe the appearance of the reaction mixture as ammonia is added, and account for the observations in terms of the ionic species present.

7. Whereas the condensation of acetylacetone with cyanoacetamide leads to a single compound in quantitative yield, reaction with cyanoacethydrazide ($NCCH_2CONHNH_2$) can lead to several condensation products, depending on the conditions. In acidic solution a compound **A** ($C_8H_9N_3O$) is obtained; mild hydrolysis of **A** gives $C_5H_8N_2$, mp 107°. In basic solution, acetylacetone plus cyanoacethydrazide give a compound **B**, isomeric with **A**. Reaction of **B** with HNO_2 gives N_2O and a compound, mp 290° ($C_8H_8N_2O$). In buffered solution, compound **C** ($C_8H_{11}N_3O_2$) is obtained; treatment of **C** with acid gives **A**. Treatment of **C** with base gives still another product, **D**, isomeric with **A** and **B**. **D** contains no CN group. Write structures for compounds **A-C** (Hint: See Table 34.1), and suggest a possible structure for **D**. [See *Angew. Chem.*, **70**, 344 (1958)].

References

Texts

R. M. Acheson, *An Introduction to the Chemistry of Heterocyclic Compounds*, 2nd Ed. Interscience, New York, 1967.

L. A. Paquette, *Principles of Modern Heterocyclic Chemistry*, W. A. Benjamin. New York, 1968.

Experimental Procedures

A. O. Fitton, and R. K. Smalley, *Practical Heterocyclic Chemistry*. Academic Press, New York, 1968.

Monographs

R. C. Elderfield (Ed.), *Heterocyclic Compounds*. John Wiley and Sons, New York, 1950–1967. (In 9 volumes.)

A. Weissberger (Ed.), *Heterocyclic Compounds*. Interscience, New York, 1950–. (Multivolume.)

CHAPTER 35

IDENTIFICATION
OF UNKNOWNS

Along with synthesis and the examination of reaction mechanisms, an equally important part of organic chemistry has to do with the characterization and identification of compounds, which may be encountered in sources ranging from a laboratory reaction to exotic tropical plants. In any case, sufficient information must be accumulated to establish the identity of the compound in question with that of a previously described compound of known structure or else to determine, *ab initio,* the structure of the unknown.

In earlier days, organic chemists relied heavily on chemical behavior in the characterization of compounds and structure elucidation. Various reactions were applied to diagnose the presence of functional groups and structural units; final evidence for a new structure usually involved systematic degradation to identifiable products. The development of spectroscopic methods has had a revolutionary effect on this area of organic chemistry. Nowadays, it is possible to characterize and arrive at the structure of a previously unknown compound entirely by physical and spectroscopic methods, without recourse to any "wet chemistry" at all. Very often, the structure stands revealed as soon as the ultraviolet, infrared, nmr, and mass spectra are in hand. With highly complex molecules, the entire structure can be determined by X-ray crystallography.

The approach to the problem of identifying or assigning the structure of an unknown organic substance will of course depend on the circumstances and the source of the sample. If the compound has been obtained as a component in a mixture of naturally occurring alkaloids or steroids, it will in all likelihood represent a variation of a known pattern, and the structural problem is relatively restricted, although subtle stereochemical differences, for example, may still present a challenging problem. Similarly, with an unknown arising as a byproduct in a synthesis, one can generally assume some relationship to the starting materials, and after a few pieces of information are obtained, a probable structure may be inferred.

Another type of situation, sometimes mentioned in textbook problems **203** but hopefully very rarely encountered in practice, is one in which the labels have come off all the bottles in the storeroom. In this case, the only premise that can be made is that most of the unknowns resulting from the disaster are to be found among the 15,000-odd entries in chemical suppliers' catalogs. Although artificial, this is essentially the context of this experiment in which unknown samples selected from the entire range of simple organic compounds are to be identified.

In the classical approach to qualitative organic analysis, the main guidance that a student had in solving an unknown was a rather rigid classification scheme and a table of compounds for each of the principal functional groups, arranged according to increasing boiling point or melting point. This approach places a great deal of emphasis on these two physical properties and requires that most or all of the unknowns be included in relatively limited tables.

The present experiment is intended to be of a more open type, with infrared and nmr spectra as well as melting or boiling points providing orientation. With spectral data available, a broader range of unknowns is possible, and the approach in some cases assumes some of the character of a structural determination rather than simply narrowing a given list of compounds to one member.

The identification process will not follow a fixed pattern and may vary considerably with different unknowns. The objective of the experiment should be to apply the methods available as efficiently as possible in arriving at a firm, well documented identification. Spectral data can and should take the place of a number of the older chemical methods for detecting functional groups, but this identification experiment is not intended to be purely an exercise in spectral interpretation. It will often be necessary to seek information from hydrolytic or oxidative degradation of the unknown. In a particular situation, some special reaction or confirmatory test may be uniquely appropriate. Frequent reference to general textbooks and library sources will be essential. Although a thorough and complete job should be done, unnecessary or irrelevant steps should be avoided so that as many unknowns as possible can be done in the time allotted; some will require much more time and effort than others.

GENERAL APPROACH

The first step in the identification process will be to obtain physical constants, infrared and nmr spectra, and solubility properties of the unknown. These data will then be assessed, and further information as needed will be obtained to permit a tentative conclusion of structure. The identification is to be completed by locating the compounds or candidate compounds in handbook tables or other literature and by preparing derivatives for confirmation.

The unknowns provided in this experiment are of the purity normally encountered in commercial organic chemicals, which is usually in the range 95 to 99 per cent. Minor impurities will generally not interfere in the identification procedure, but preliminary recrystallization or distillation of a sample of the unknown may be desirable.

PHYSICAL PROPERTIES

For solid unknowns, the melting point is determined in the usual way, raising the temperature of an initial sample rapidly to get an approximate range, then repeating at a rate of 2 to 3° per minute in this region. If a sample is recrystallized, the melting point should be checked before and after.

To determine the boiling point of a liquid unknown, place 1 ml of the liquid in a 18 × 150 mm test tube equipped for reflux with a coil of rubber tubing (Fig. 12.1). Suspend a thermometer with the bulb below the level of the tubing and add a boiling stone. Heat with a low flame until gentle refluxing occurs on the walls and the thermometer. When a constant temperature is registered on the thermometer, record this value as the boiling point.

Alternatively, a few ml of the liquid can be distilled in a small flask (Fig. 35.1). This set-up must be used for highly volatile compounds; if in doubt consult your instructor. The flask is mounted on an asbestos board with a small hole so that excessive heating of the glass is avoided; a chilled test tube fitted over the side arm serves as a receiver. With these small samples, care must be taken to distill enough material to obtain the equilibrium boiling point.

A few of the unknowns cannot be distilled at atmospheric pressure without some decomposition; for these compounds the boiling point is of

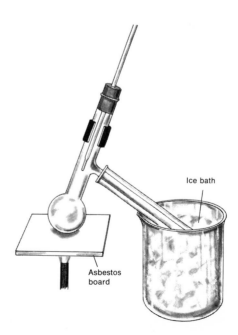

Ice bath

Asbestos board

FIGURE 35.1 Distillation of unknowns.

little use, since the literature value will usually be recorded at some reduced pressure, and in such cases, it will be stated that distillation of the
unknown should not be attempted.

The melting point of a solid or the boiling point of a liquid is tradi-
tionally one of the first physical constants cited in characterizing an organic
compound; in early work these temperatures were among the few measurements that could be made. Several other properties, such as the refractive
index and density of liquids and optical rotation of disymmetric compounds, are often recorded for additional characterization, although these
have become less important since the advent of routine spectral data, and
they are not particularly useful in locating a compound in the literature.

In using the observed melting point or boiling point of an unknown or
a derivative for comparison with literature values, it is necessary to allow
sufficient "leeway" and to consider a range of several degrees in literature
values on either side of the observed temperature. Literature values as well
as the observed ones are subject to inaccuracies; frequently more than one
value can be found because of differences between individual investigators.

The melting point of a compound conveys little structural information
in itself, since it depends on such diverse factors as molecular size, symmetry, rigidity, and polarity of functional groups. Although there is a regular progression of melting points within a typical aliphatic homologous
series, large disparities in the melting points of closely similar compounds
can arise due to differences in molecular shape, as illustrated with the
polyols erythritol and pentaerythritol.

$$\begin{array}{cc}
\overset{\displaystyle OH \quad OH}{\underset{\displaystyle HOCH_2-CH-CH-CH_2OH}{\big| \quad \big|}} & \overset{\displaystyle CH_2OH}{\underset{\displaystyle \underset{\displaystyle CH_2OH}{\big|}}{HOCH_2-C-CH_2OH}}
\end{array}$$

Erythritol, mp 121°   Pentaerythritol, mp 254°

The boiling point of a liquid is much more directly related to the functional group and molecular size, since forces operating in the liquid state
are less affected by symmetry and rigidity than those in a crystal. Within
any straight-chain aliphatic homologous series, the boiling point increases
in a regular way with increasing molecular weight, with increments between successive members becoming smaller, the longer the chain.
Branching, particularly in the vicinity of a functional group, markedly
lowers the boiling point within a group of isomers. The presence of an
alicyclic or aromatic ring causes a significant increase in boiling point over
that of an aliphatic compound having the same functional groups and
number of carbon atoms. Table 35.1 contains representative data on the
boiling points of a few simple monofunctional aliphatic alcohols; within
each group, the lowest boiling point is that of the most highly branched
isomer, and the highest boiling point corresponds to the longest chains
(*n*-alkanol).

TABLE 35.1 BOILING POINT RANGES OF ALIPHATIC ALCOHOLS

TYPE	4-CARBON	6-CARBON	8-CARBON
Primary	108–116°	148–156°	183–194°
Secondary	99°	120–139°	150–180°
Tertiary	83°	120–125°	149–165°

It is obvious that a relatively low boiling point for an unknown, say below 100°, greatly limits the number of possible compounds, and may even define the structure uniquely with little other data. A boiling point in the range 100 to 160°, together with information on functional groups, provides a rough indication of molecular size. On the other hand, the number of possible compounds increases enormously in this range, and the boiling point cannot be used to pinpoint one or two candidates if all known compounds are admitted as possibilities.

SOLUBILITY CLASSIFICATION

A good deal can be learned about a compound from its solubility in a few media; the solubility classification complements spectral data and helps to determine the direction of further work. The solubility of the unknown is checked in water, dilute acid, dilute base, and concentrated sulfuric acid in that order, stopping when a positive result is obtained. If the compound is soluble in water, nothing is learned by testing in dilute acid or base; if it is soluble in any aqueous medium, no information is gained from solubility in concentrated H_2SO_4.

Water. Solubility of an organic compound in water reveals the presence of ionic or "polar" groups which can be solvated or can participate in hydrogen bonding. The extent of solubilization depends, of course, on the "ratio" of functional group to carbon skeleton. Liquids containing hydroxyl, carboxyl, amino, or amide groups and no more than four or five carbon atoms are miscible with water in all proportions (indicated by "∞" in solubility tables) or have appreciable solubility. With two or more of these groups, a much larger molecule will be water-soluble.

With solids, crystal structure has a major influence, and solubility in any solvent, water or organic, is related to melting point as well as the polarity of the molecule. Thus oxamide, $NH_2COCONH_2$, which has the unusually high melting point of 410°, is very sparingly soluble in water.

In testing for solubility in water, about 50 mg of solid or 0.1 ml of liquid is added to 1 ml of water in a 10 × 75 mm test tube. Hard dense crystals of a solid may dissolve slowly and should be powdered and stirred well. If a clear solution is obtained, or a major amount of the compound dissolves with the amounts specified, the compound is considered "soluble" in water.

If the unknown is soluble in water, the pH should be estimated with indicator paper to detect the presence of acidic or basic groups in the molecule. The solubility in ether should also be checked if the unknown is readily soluble in water. A multiplicity of polar groups, as in a polyol, may render the compound insoluble in ether. With a solid, solubility in water and insolubility in ether suggests the possibility of an ionic salt, and this should be explored by treating the solution with acid or base.

Aqueous Acid or Base. Compounds that can be converted to ionic species can be recognized by the solubility in water at certain pH values. Three major classes of compounds can be distinguished in this way: *strong acids, weak acids*, and *bases*. All compounds with $K_a > 10^{-12}$ ($pK_a < 12$) are converted to anions to a significant extent in $1N$ NaOH (pH 13). Acids with pK_a in the range 3–7, including all carboxylic acids and nitrophenols, are also soluble in aqueous sodium bicarbonate. Solubility in NaOH, but not $NaHCO_3$, indicates a weak acid such as a phenol, enol, or aliphatic nitro compound. Organic bases with K_b $10^{-3} - 10^{-10}$ are soluble in $1N$ HCl by virtue of the conversion to cations. The only compounds which can be protonated in dilute aqueous acid are amines.

$$RCO_2H + HCO_3^- \rightarrow RCO_2^- + CO_2 + H_2O$$
$$R_3N + H_3O^+ \rightarrow R_3NH^+ + H_2O$$

In detecting basic or acidic properties by these solubility criteria, the important point is whether the unknown is significantly more soluble in aqueous acid or base than in water. If a test is doubtful, neutralization of the test solution should cause reprecipitation of an amine or an acid which is dissolved due to salt formation. One possible complication that must be kept in mind is the relatively low solubility of certain salts, particularly in the presence of an excess of the common ion. For example, it is possible to mistake the formation of a sparingly soluble hydrochloride for insolubility of a solid amine.

Concentrated Sulfuric Acid. Virtually all compounds of moderate molecular size that contain a nitrogen or oxygen atom or a double or triple bond are protonated in 96% H_2SO_4, and therefore dissolve to some extent. This test should be deferred until after the infrared spectrum is obtained. If there is no clear indication of any functional group in the spectrum, it will then be useful to check the solubility in concentrated H_2SO_4 to detect the presence of ether oxygen, reactive double bonds, and so forth. The solution may become dark in color, or polymer may separate; these reactions constitute a positive test, but slight darkening, without actual solution, may be due to trace impurities.

SPECTRA AND ANALYTICAL DATA

The two most widely useful types of spectra, infrared and nmr, will be obtained for each unknown. These spectra are not to be considered as extra data to be used if chemical analysis is unsuccessful; rather they should be the first data analyzed and should serve as the basis for deciding what additional steps are appropriate and necessary. Interpretation of both the nmr and infrared spectra should be included in your report of the analysis of an unknown. For general information on obtaining and interpreting these spectra refer to the sections on spectral methods (Chapts. 9 and 10).

In certain cases you may be given additional information about an unknown in the form of an elemental analysis (% C, H, N, and so forth) or a molecular weight. The latter, if determined by measuring accurately a mass spectrum, can fix the molecular formula. Approximate ($\pm 10\%$) values, such as would be obtained by freezing point determinations, are useful only in conjunction with other data.

In using an elemental analysis, it must be recognized that methods of analysis are such that the results are accurate to only $\pm 0.3\%$ (absolute), and that an empirical and not necessarily a molecular formula is obtained. As an example, consider a compound containing C, H, and possibly O, which analyzed for 74.02% C and 6.48% H. By difference it contains 19.5% O. The calculated molar proportions of C, H, and O are thus 5.06:5.27:1. Rounding to $C_5H_5O_1$, the theoretical analysis is 74.06% C and 6.21% H, in good agreement with that found. Since a compound containing only C, H, and O cannot have an odd number of hydrogen atoms, the molecular formula must be $(C_{10}H_{10}O_2)_x$. The value of x can usually be determined by considering other data.

On the following pages, three examples of typical unknowns are given, including data on physical properties and solubility, and reproductions of infrared and nmr spectra (Figs. 35.2, 35.3, and 35.4). The discussions which follow the data illustrate the approach which should be followed in their interpretation and the conclusions that can be reached.

EXAMPLE 1

A liquid, bp 225–235°, insoluble in water and neutral to aqueous acid or base, but soluble in concentrated H_2SO_4.

1. Infrared Spectrum

a. Broad strong band at 3360 cm^{-1} indicates bonded OH group typical of alcohols run in liquid film.

b. Bands at 3050 and at 2880–2950 suggest aryl or olefinic and aliphatic C—H, respectively.

c. Absence of strong bands in 1800–1650 region indicates no carbonyl groups; 1600 and 1500 cm^{-1} bands are typical of aromatic ring.

d. Band at 1460 cm^{-1} is due to aliphatic C—H bending.

e. Strong absorption at 1030–1050 cm^{-1} is in region of C—O stretching of alcohols and confirms an OH group.

f. Sharp bands at 700 and 740 cm^{-1} are very characteristic of a benzene ring with one substituent.

2. Nmr Spectrum

INTEGRATION. The spectrum contains five groups of signals with relative areas, reading upfield (from left to right), of $7:2.9:1.5:2.8:2.8$ (these are the vertical heights measured in cm on the original chart). Dividing by 1.4 leads to an integer ratio of $5:2:1:2:2$ for the five peaks.

CHEMICAL SHIFTS

a. The highest field multiplet at 1.83 ppm is within the range of either CH_3, CH_2, or CH groups in various environments, but the fact that this and the next two multiplets each represent two protons strongly suggests three CH_2 groups with different combinations of substituents. Also, the complex splitting pattern rules out a CH_3 group; any CH_3 group below 1.7 ppm would give either a singlet or a doublet, since it would have to be present in a grouping CH_3X or CH_3CHX_2.

b. The multiplet at 2.62 ppm requires a substituent such as a double bond or an aromatic ring; the latter is an obvious probability from other evidence.

c. The position of the most deshielded multiplet at 3.54 ppm requires an adjacent electronegative atom such as oxygen or halogen. Moreover, this peak is poorly resolved, suggesting coupling with a slowly exchanging neighboring OH proton.

d. The singlet at 3.1 ppm indicates an OH proton, which can be confirmed by exchange with D_2O.

e. The sharp 5-proton singlet at 7.12 ppm clearly indicates a C_6H_5 grouping.

3. Analysis

Two major structural features that stand out in both spectra are C_6H_5 and OH groups. Lack of solubility in base confirms that the OH is not located on the ring. The rest of the molecule appears from the nmr spectrum to consist of three CH_2 groups, and the only construction that can be placed on this combination is a $CH_2CH_2CH_2$ chain, linking C_6H_5

and OH. This arrangement is exactly that inferred from the chemical shifts of the three CH_2 multiplets, i.e.,

$$\underset{C_6H_5-CH_2-CH_2-CH_2-OH}{\underbrace{2.62}\quad\underbrace{1.83}\quad\underbrace{3.54}}$$

This structure is consistent with the splitting patterns of the CH_2 multiplets. The patterns are complex and cannot be analyzed exactly from this spectrum, but the highest field multiplet (1.83 ppm), which is assigned to the central CH_2 group on the basis of the chemical shift, has the largest number of neighboring protons and is the most highly coupled.

We therefore reach the tentative conclusion that the compound is 3-phenylpropanol; the boiling point of this compound is recorded as 237°, in the general range observed. This identification would be confirmed by the preparation of a derivative (p. 224). Of several possibilities reported, the 3,5-dinitrobenzoate ester (mp 92°) is convenient and has a suitable melting point.

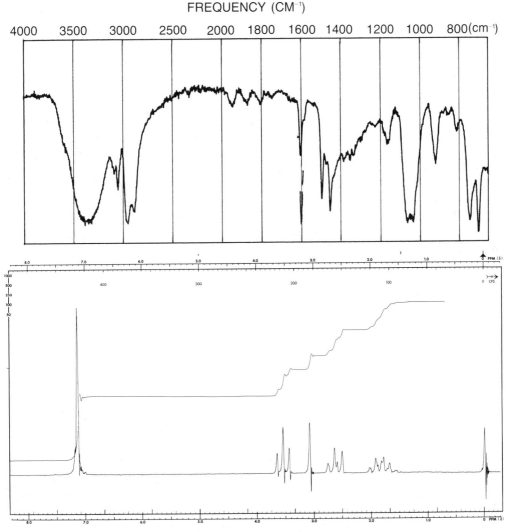

FREQUENCY (CM⁻¹)

FIGURE 35.2 Spectra for Example 1.

EXAMPLE 2

A liquid, bp 101–103°, slightly soluble in H_2O, neutral toward 5% HCl and 5% NaOH.

1. Infrared Spectrum

a. Weak bands at 2960 and 3100 cm^{-1} suggest both aliphatic C—H and =C—H (either olefinic or aromatic), respectively. No OH or NH peaks are apparent in the 4000–3200 cm^{-1} region.

b. The strong band at 1740 cm^{-1} indicates a C=O group; the strong 1230 cm^{-1} band can be assigned to C—O—C stretching, which taken in conjunction with the 1740 cm^{-1} band suggests the possibility of an ester.

c. The weak band at 1650 cm^{-1} suggests C=C; probably not conjugated because of low absorbance.

2. Nmr Spectrum

INTEGRATION. The spectrum consists of four separate groups of signals. The relative areas from the integration curve, from left to right, are 20:36:33:56 or 1:1.8:1.65:2.8. Dividing by 18 rather than 20 gives 1.10:1.97:1.82:3.09 which is within the 5 to 10 per cent error of this measurement for 1:2:2:3.

CHEMICAL SHIFTS

a. The singlet at 2.06 ppm (3 protons) clearly indicates a CH$_3$ group probably adjacent to a carbonyl.

b. The doublet with small additional splitting at 4.57 ppm (2 protons) suggests a CH$_2$ group flanked by two deshielding atoms or groups.

c. Multiplets in the range 5 to 6.5 ppm (3 protons total) can be assigned to olefinic protons (directly attached to C=C). The simplest possibility, consistent with the extensive splitting, is a vinyl (—CH=CH$_2$) group.

3. Analysis

Taking the basic conclusions drawn (carbonyl, vinyl, methyl and methylene groups) without any of the qualifications given, several structures might be suggested:

1. $CH_3CH_2COCH=CH_2$
2. $CH_3COCH_2CH=CH_2$
3. $CH_3OCOCH_2OCH=CH_2$
4. $CH_3COCH_2CH=CH_2$
5. $CH_3CO_2CH_2CH=CH_2$

Structure 1, ethyl vinyl ketone, contains these groups and its boiling point (102°) agrees well with that of the unknown. However, this structure can be eliminated at once, since the CH$_3$ protons appear as a singlet and not a triplet.

The splitting pattern in the nmr spectrum is satisfied by allylmethyl ketone (2, bp 105°), but the chemical shift of the CH$_2$ protons (4.57 ppm) is at a significantly lower field than expected for a —COCH$_2$C= group ($\delta \sim 3.25$ ppm—see Table 10.2, p. 88). Moreover, the infrared band for C—O—C is not accounted for by either structure 1 or 2.

Insertion of an oxygen atom between the CH_2 and $CH{=}CH_2$ groups, as in structure 3, provides for this infrared absorption and the low-field position of the CH_2 signal; the splitting of the CH_2 is inconsistent with 3, however.

These three ketone structures could be eliminated on chemical grounds by means of a 2,4-dinitrophenylhydrazine test on the unknown, which does not give a positive reaction for hydrazone formation.

Of the two esters, methyl 3-butenoate (4, bp 108°) is ruled out by the chemical shift of the CH_3 signal as well as that of the CH_2. The remaining possibility, allyl acetate (5, bp 104°) satisfies all of the spectral data. Confirmatory evidence for this structure could be obtained by saponification, although isolation of the products, which would be volatile and water-soluble, could not be easily done on a small scale. Derivatives such as an amide of the acid portion could be obtained directly from the ester (p. 226). Another possibility in this case would be synthesis of an authentic sample of the ester and comparison of infrared spectra.

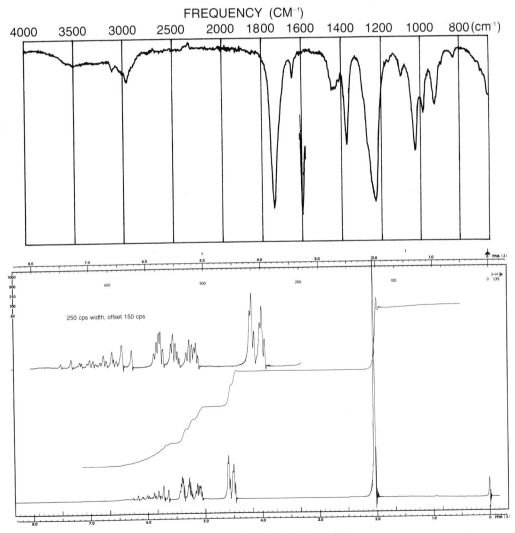

FIGURE 35.3 Spectra for Example 2.

EXAMPLE 3

A solid, mp 84°, insoluble in water, dilute acid or aqueous bicarbonate, but appreciably soluble in aqueous NaOH.

1. Infrared Spectrum

a. The band at 3300 cm^{-1} is in the range of OH and NH stretching; the sharpness of this peak is inconsistent with OH and amine or amide NH$_2$ bands in solid-state spectra, and suggests the NH band in a mono-substituted amide, RCONHR.

b. Bands at 2950 and 3040 cm^{-1} indicate both alkyl and aromatic or olefinic CH.

c. Strong sharp bands at 1705 and 1670 cm^{-1} reveal two carbonyl groups; the former probably a ketone, and the latter possibly an amide.

d. The sharp 1595 cm^{-1} band is presumably due to aromatic ring.

e. The strong band at 1530 cm^{-1} when taken together with NH and CO assignments can be attributed to the "amide II" combination and is a further indication of the NHCO grouping.

f. Of the remaining sharp peaks, the only one that can be assigned is the strong 755 cm^{-1} band which indicates a benzene ring with four adjacent hydrogen atoms.

2. Nmr Spectrum

INTEGRATION. The spectrum contains six main peaks with an area ratio of 0.9 : 1.1 : 3.0 : 3.0 : 1.9 : 2.8 which reduces to 1 : 1 : 3 : 3 : 2 : 3, or a total of 13 protons. There are also two very small peaks (1.90 and 5.00 ppm) which are distinct from side bands and suggest the presence of a small amount of an impurity or a second compound.

CHEMICAL SHIFTS

a. The 3-proton singlet at 2.28 ppm indicates CH$_3$ adjacent to C=O.

b. The 2-proton singlet is evidently due to CH$_2$ adjacent to at least one and possibly two deshielding substituents, such as C=O or aryl.

c. The 3-proton singlet at 3.87 must arise from a CH$_3$O group, and the position indicates that it is attached to an aromatic ring.

d. The 3-proton multiplet at 7.0 ppm suggests three protons with slightly different electronic influences on a benzene ring.

e. The multiplet at 8.25 ppm (upper offset trace) appears to represent a fourth, highly deshielded aryl proton.

f. The broadened peak at 9.1 ppm is characteristic in both shape and position for an amide NH.

3. Analysis

From the solubility data, the compound is a weak acid, probably phenolic or enolic; this inference could be confirmed independently by a color test with ferric chloride. Neither the infrared nor the nmr spectrum indicates an OH group, however.

Taking stock of the data, various lines of evidence establish the presence of the following components: (1) a benzene ring with four (adjacent) hydrogens, (2) o-CH$_3$OC$_6$H$_4$, (3) CH$_3$CO, (4) NHCO, accounting for 11 of the 13 protons in the nmr spectrum. The deshielded CH_2 group remains to be located, and the other groups must be connected in some way. There are four possibilities, as follows:

1. $o\text{-}CH_3OC_6H_4\text{---}CH_2\text{---}CONH\text{---}COCH_3$
2. $o\text{-}CH_3OC_6H_4\text{---}CH_2\text{---}NHCO\text{---}COCH_3$
3. $o\text{-}CH_3OC_6H_4\text{---}CONH\text{---}CH_2\text{---}COCH_3$
4. $o\text{-}CH_3OC_6H_4\text{---}NHCO\text{---}CH_2\text{---}COCH_3$

Of these alternatives, the diacylimide (**1**) and the β-ketoamide (**4**) could both conceivably have weakly acidic character. A distinction could readily be made by chemical means, since hydrolysis would lead to completely different types of compounds. The imide (**1**) on treatment with base would give o-methoxyphenylacetic acid, ammonia and acetic acid, whereas (**4**) would be hydrolyzed to o-anisidine.

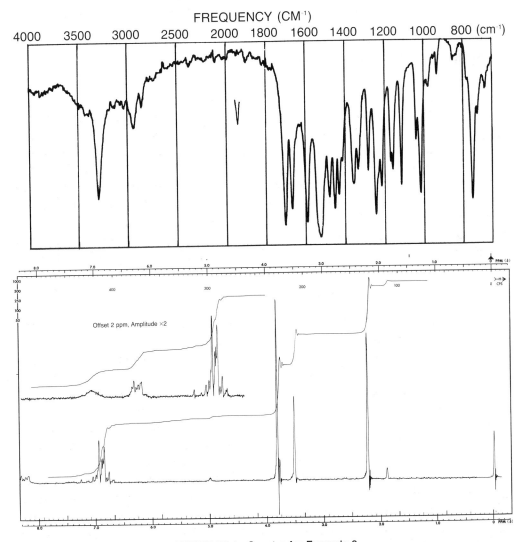

FIGURE 35.4 Spectra for Example 3.

1. $o\text{-}CH_3OC_6H_4CH_2CONHCOCH_3 \rightarrow o\text{-}CH_3OC_6H_4CH_2CO_2H$
$$+ NH_3 + CH_3CO_2H$$

4. $o\text{-}CH_3OC_6H_4NHCOCH_2COCH_3 \rightarrow o\text{-}CH_3OC_6H_4NH_2 + 2\ CH_3CO_2H$

In the event, refluxing the compound with aqueous NaOH gave an oily amine and not ammonia, indicating the β-ketoamide (4). The melting point of o-acetoacetanisidide is found in tables to be 83-85°, in agreement with that observed for the unknown. A derivative can be readily obtained by acetylation of the anisidine formed on hydrolysis.

A further point in the nmr spectrum is of significance in the light of the β-ketoamide structure. The two small peaks at 1.90 and 5.00 ppm are in the positions expected for the methyl and methine protons, respectively, of the enol form $o\text{-}CH_3OC_6H_4NHCOCH{=}C(OH)CH_3$. The integral of the 1.90 ppm peak matches the slight deficiency in the CH_3CO integral (2.8 vs 3.0 for the other 3-proton signals) and suggests about 5 per cent of the enol. Such a small amount would not give recognizable OH peaks with the recording conditions used for either the infrared or nmr spectrum.

ADDITIONAL DATA ON UNKNOWNS

After assessing the information obtained from physical properties, solubility, and spectra, as illustrated in the preceding pages, it will frequently be possible to draw a tentative conclusion as to the identity of the unknown and proceed to selection and preparation of a derivative. In other cases, certain features will be established, but further data will be desirable to define the environment, or even the nature of the functional group, or to decide between two possible interpretations. Additional information can usually be obtained from one or more of the approaches discussed in the following paragraphs. More complete information on these and other procedures can be obtained from the references at the end of the chapter.

DETECTION OF OTHER ELEMENTS

One important item that may be needed is information on the presence of elements other than C, H, and O. Although C—Cl bonds can sometimes be seen in the infrared spectrum (800–600 cm^{-1}), these bands may be obscured; C—Br and C—I stretching bands occur in the far infrared (< 600 cm^{-1}). A qualitative test for halogen may therefore be indicated, particularly if the nmr spectrum suggests an odd number of protons. The common functional groups containing nitrogen should be revealed by the basicity (amines) or the infrared spectrum (amide CO, C$\equiv$N, NO$_2$), but a confirmatory test for nitrogen can readily be carried out.

Detection of N, Cl, Br, and I is accomplished by the total decomposition of the compound with hot metallic sodium, followed by detection of anions in the usual way. With rare exceptions, any compound containing a C—N bond will give cyanide ion under these conditions; excess carbon is usually converted to the amorphous element.

$$[C, H, O, N, X] \xrightarrow{Na} \xrightarrow{H_2O} C, OH^-, CN^-, X^-$$

In this and other tests which require observation of a positive or negative result, there is one all-important rule: *always run a control.* It is quite futile to attempt a conclusion from the reaction of an unknown when the observer does not know exactly the appearance of a positive and a negative result. Run a known compound first, on the scale that will be used for the unknown, to see the behavior and to insure that you are doing it properly. It may be equally important in some cases to run a *blank* in order to observe a negative result. If there is doubt about the result with a known, clear it up before turning to the unknown.

Procedure. In a clean dry 10×75 mm test tube place a small piece of sodium (a cube about 4 to 5 mm on edge). Handle with a spatula tip and blot dry with filter paper. Wipe the *outside* of the test tube and do not handle with fingers. Have ready a sample of about 100 mg of a solid unknown, or 0.1 ml of a liquid in a pipette, and a small beaker containing about 10 ml of distilled water. Heat the sodium over a flame until it melts and begins to glow red. Add the sample so that it falls directly on the molten sodium and not the wall where it may be volatilized. With a very volatile liquid, add a second 0.1 ml sample. Heat the tube briefly again and then drop it, hot, into the distilled water. There may be a small explosion and spurt of flame, but this is not a hazardous operation with the amount of sodium specified. Break up the glass and dark char with a stirring rod and filter the solution through paper. If the solution is brown in color (indicating incomplete decomposition), add a little charcoal before filtering.

TEST FOR N. To 1 ml of the solution add 2 drops of saturated ferrous ammonium sulfate solution and 2 drops of 30 per cent potassium fluoride solution. Boil the mixture for 30 seconds and acidify by dropwise addition of 30 per cent sulfuric acid until the iron hydroxide just dissolves. The appearance of a brilliant blue precipitate of Prussian blue indicates the presence of cyanide ion and nitrogen in the compound.

TEST FOR HALOGEN. Acidify a 2-ml portion of the alkaline fusion solution with dilute nitric acid and boil briefly to expel any HCN that may be present. The appearance of a distinct white or yellow precipitate on addition of silver nitrate indicates the presence of halogen. A yellow color suggests bromide or iodide.

To differentiate the halogens in the event of a positive silver halide precipitate, acidify another 2- to 3-ml portion of the original solution and add about 0.3 ml of carbon tetrachloride. Then add a few drops of fresh chlorine water or a few mg of calcium hypochlorite (be sure the pH remains acidic). A yellow-orange color in the CCl_4 layer indicates bromine; a violet color, iodine.

EQUIVALENT WEIGHTS

There are many methods for quantitative determination of various functional groups in organic compounds; these reveal the number of such groups in the molecule if the molecular formula is known. If the formula of the compound is not known, quantitative analysis of a particular group provides the *equivalent weight,* i.e., the number of grams of compound containing one group.

The simplest of all these quantitative methods is acid-base titration, which is used to determine the *neutralization equivalent* of acids. Practically any carboxylic acid can be titrated with standard base to a sharp endpoint with phenolphthalein, since the pH at complete neutralization for an acid of pK_A 3–6 is somewhat above 7. With care, the neutralization equivalent of a pure, dry acid can easily be determined with an accuracy of 1 per cent, providing very valuable information about an unknown acid. The neutralization equivalent of a monobasic acid is the molecular weight; for di- or polybasic acids, it is some integral fraction of the molecular weight. The procedure given can also be used for the neutralization equivalent of amine salts, but it is generally difficult to obtain the latter sufficiently pure and dry.

Procedure. Weigh out a 150 to 200 mg sample of benzoic acid to ±1 mg on an analytical balance. Place the acid in a 125 ml Erlenmeyer flask with 50 ml of water, add 2 to 3 drops of phenolphthalein solution, and titrate to a pink endpoint with ~0.1N NaOH from a 25 or 50 ml buret. From the volume of titrant and weight of benzoic acid (eq. wt. 122) used, calculate the exact normality of the NaOH solution. As a check of your technique and the result obtained, titrate a second sample of benzoic or another known acid.

Repeat the process using the standardized base with a similar accurately measured quantity of the unknown acid (*in duplicate*), and calculate its equivalent weight. Liquid acids should be freshly distilled and weighed in a capped vial with minimum exposure to air since they are generally hygroscopic. Solid unknown acids should be recrystallized and thoroughly dried. Acids which are very insoluble in water can be titrated in aqueous alcohol. If this is necessary, the standardization should be carried out with the same amount of added alcohol.

Other quantitative methods, including the saponification equivalent of an ester or quantitative hydrogenation of an olefinic compound, may be desirable in certain cases. Procedures for these or other special methods should be obtained from the literature, and arrangement made with the instructor for necessary reagents and equipment.

CHEMICAL TRANSFORMATIONS

Reactions of several kinds may be required to convert the unknown to a simpler compound or to recognizable fragments. A few of the general approaches are mentioned here; additional possibilities may be suggested by spectral data and other observations.

Hydrolysis. Esters, anhydrides, amides, and nitriles can usually be hydrolyzed in aqueous or alcoholic base; hydrolysis with fairly concentrated acid may be needed for certain amides. The acid or other hydrolysis products can often serve as derivatives, and typical procedures are given in the following section.

Acetals and related carbonyl derivatives such as enol ethers or enamines are susceptible to hydrolysis in acid, and if the presence of these groupings is suspected, hydrolysis in warm dilute aqueous or alcoholic acid will liberate the aldehyde or ketone which can be isolated or characterized as a derivative. The conditions suggested will usually open an epoxide ring and can also lead to dehydration; these possibilities must be kept in mind.

$$\begin{array}{c} \text{OR}' \\ \diagdown\diagup \\ \diagup\mathrm{C}\diagdown \\ \text{OR}' \end{array} \xrightarrow{\text{H}_3\text{O}^+} \begin{array}{c} \diagdown \\ \diagup \end{array}\mathrm{C}{=}\mathrm{O} + 2\ \text{R}'\text{OH}$$

$$\begin{array}{c} \diagdown \diagup \\ \mathrm{C}{=}\mathrm{C} \\ \diagup \diagdown \\ \text{OR} \end{array} \xrightarrow{\text{H}_3\text{O}^+} \begin{array}{c} \diagdown \\ \diagup \end{array}\mathrm{CH}{-}\overset{|}{\mathrm{C}}{=}\mathrm{O} + \text{R}'\text{OH}$$

Enolizable β-dicarbonyl compounds, whose presence is usually revealed by a red or violet color with ferric chloride, can be cleaved by base to give ketone or acid products.

$$\mathrm{RCOCH_2COR'} \xrightarrow{\text{OH}^-} \mathrm{RCO_2H} + \mathrm{CH_3COR'}$$

$$\mathrm{RCOCH_2CO_2Et} \xrightarrow{\text{OH}^-} \mathrm{RCO_2H} + \mathrm{CH_3CO_2H} + \text{EtOH}$$

Treatment with aqueous or alcoholic alkali may also bring about aldol condensation and occasionally reverse aldol or Michael reactions. Although the latter reactions will more often be encountered unexpectedly rather than by design, they can be highly informative when correctly interpreted.

It is clear that numerous functional systems may respond to hydrolytic treatment, and if an impasse is reached in pursuing an unknown, hydrolysis will sometimes turn up a lead. When probing for a reaction in this way, the first observation to be made is simply whether reaction has occurred, i.e., whether the compound has been consumed or can be recovered unchanged.

Oxidation. Numerous oxidation methods have been developed for various purposes in organic chemistry. These range from highly selective reagents such as manganese dioxide for conversion of allylic alcohols to aldehydes, to rather drastic procedures for degrading carbon chains. A few of the oxidations that may be useful in this experiment are mentioned here, but specific procedures are not given since they vary with the reagent and compound. Directions must always be obtained from a reference at the end of the chapter or some other source before undertaking an oxidation procedure on an unknown.

Chromic acid at room temperature converts primary alcohols and aldehydes to acids, and secondary alcohols to ketones (see Chapt. 20). The latter transformation is often useful in dealing with an aliphatic alcohol, since derivatives of the ketone may be more reliable. Double bonds may undergo oxidation with cold chromic acid, leading to carboxylic acid or ketonic fragments.

More vigorous treatment with hot chromic acid or alkaline permanganate can effect complete oxidation of a side chain in an aromatic hydrocarbon or a nitro or halogen derivative to an acid with the carboxyl group adjacent to the ring. These reagents will of course oxidize any more susceptible grouping very rapidly.

The iodoform reaction provides a selective method for transformation of a methyl ketone or methyl carbinol to an acid; a yellow precipitate of iodoform is confirmatory evidence for these groups.

$$I_2 + NaOH \rightarrow NaI + NaOI$$
$$CH_3CHOHR + NaOI \rightarrow CH_3COR + H_2O + NaI$$
$$CH_3COR + 3\ NaOI \rightarrow CI_3COR + 3\ NaOH$$
$$CI_3COR + NaOH \rightarrow CHI_3 + RCO_2Na$$

Glycol or ketol groups are cleaved selectively by periodic acid to carbonyl or acid fragments; the iodic acid formed can be detected as a confirmatory test by precipitation with silver nitrate.

$$\underset{\underset{OH}{|}}{RCH}—\underset{\underset{OH}{|}}{CH}—R' + HIO_4 \longrightarrow RCHO + R'CHO + HIO_3 + H_2O$$

$$Ag^+ \searrow$$
$$AgIO_3$$

DERIVATIVES

The final point in the identification is conversion of the unknown to a solid derivative whose melting point can be compared with a literature value. The derivative may be any compound that is formed in a reaction which is characteristic for the unknown. It is obviously desirable to choose, when possible, a derivative that is obtained in good yield and readily isolated and has a melting point in the most convenient region, i.e., 80 to 180°. In many cases, the derivative will be confirmatory evidence for a tentative conclusion, but occasionally it may be necessary to choose between two possible candidates; in this case, the derivative or reaction product from both compounds must be known.

For acids, alcohols, amines, and carbonyl compounds, a variety of simple condensation products can serve as derivatives, and Handbook tables contain quite complete listings. Procedures for a few of the more useful of these standard derivatives are given below. Solid acids obtained by hydrolysis of esters, amides, or nitriles are usually satisfactory derivatives if the acid represents the major portion of the molecule. It may, however, be desirable to characterize the entire molecule by a more specific reaction. For example, hydrolysis of acetanilide gives acetic acid and aniline, both liquids. Either the acid or amine or both could be isolated and converted to solids, but a much simpler and far more effective derivative is obtained by bromination to p-bromoacetanilide.

A number of the unknowns may lend themselves to special derivatives which will be suggested by the tentative structure deduced from spectral and other data. Cyclization products can be readily obtained from many bifunctional compounds, and occasionally rearrangement or partial degradation will provide highly characteristic derivatives. In such cases, details for carrying out the reaction and isolation of the product should be obtained from the literature for the specific compound that is under consideration.

ACIDS

The most satisfactory general derivatives of acids are amides, particularly the anilides and p-toluides, which are prepared by the general sequence:

$$RCO_2H + SOCl_2 \rightarrow RCOCl + SO_2 + HCl$$

$$RCOCl + 2\ R'NH_2 \rightarrow RCONHR' + R'NH_3^+Cl^-$$

The second reaction requires the use of some base to combine with HCl. In the derivatization of an acid, an excess of the amine is usually used, but if the amine is the important component, some other base, usually aqueous NaOH, is used as the acid acceptor.

It should be noted that in the reaction with thionyl chloride, polyfunctional acids will often give cyclization or condensation products rather than an acid chloride. Thus dibasic acids may form anhydrides, and α-acylamino acids give oxazolones (azlactones).

Procedure

ACID CHLORIDE. Mix about 1 g or 1 ml of the acid with 2 ml of thionyl chloride and reflux gently for 15 minutes. (Care must be taken to use an extremely small flame to avoid over-heating the wall of the flask above the liquid level and charring the contents.)

For conversion to the anilide or p-toluidide, dilute the acid chloride with 5 to 10 ml of methylene chloride, and add it to a solution of 2 g of aniline or *p*-toluidine in 10 ml of methylene chloride. After mixing, allow the reaction to stand for a few minutes, add 10 ml of water and transfer the mixture to a separatory funnel. Add more solvent if necessary to dissolve all of the amide. At this point it is convenient to add sufficient ether to make the organic layer lighter than water. Wash the organic phase with dilute HCl until the aqueous layer is acidic, then with bicarbonate solution and finally water. Dry the organic phase (MgSO$_4$) and evaporate on the steam bath and crystallize and collect the derivative.

Conversion to the amide may be preferable if the acid is a high melting solid. In this case, add the acid chloride, without dilution, *dropwise* to a mixture of 10 ml of concentrated ammonium hydroxide and an equal volume of ice.

ALCOHOLS

The most generally useful derivatives of alcohols are esters of substituted benzoic or carbamic acids; the carbamate esters are commonly called urethanes. *p*-Nitrobenzoates or 3,5-dinitrobenzoates are obtained by treatment of the alcohol with the acid chloride and pyridine, which serves as a catalyst and acid acceptor.

$$ArCOCl + ROH + C_5H_5N \rightarrow ArCO_2R + C_5H_5NH^+Cl^-$$

Urethanes are prepared from the alcohol and an aryl isocyanate;

$$ArN{=}C{=}O + ROH \rightarrow ArNHCO_2R$$

In any reactions with isocyanates, the following reactions leading to the diarylurea will occur if water is present.

$$ArN{=}C{=}O + H_2O \rightarrow [ArNHCO_2H] \rightarrow ArNH_2 + CO_2$$

$$ArN{=}C{=}O + ArNH_2 \rightarrow ArNHCONHAr$$

The urea is a high-melting insoluble compound (diphenylurea, mp 238°) and can seriously interfere with the isolation of the desired derivative. Alcohols should be dry, and an excess of isocyanate must be avoided.

Procedures

NITROBENZOATES.   Dissolve 1 ml of the alcohol in 3 ml of pyridine and add 0.5 g of the acid chloride. Warm the solution for a few minutes and pour into 10 ml of water. If a well crystallized ester separates, this is collected; otherwise, extract the product with ether and wash the ether solution free of pyridine with dilute HCl, dry, and evaporate.

URETHANES.   To about 0.5 ml or 0.5 g of the anhydrous alcohol add 0.2 to 0.3 ml of phenyl or naphthylisocyanate and warm the solution for a few minutes (do not expose to steambath vapors). Evaporate the solution to a syrup; a typical crystallization procedure is given on page 19. If a very sparingly soluble precipitate separates, it is probably the urea; this must be removed by filtration, followed by evaporation, dissolving in a small volume of ether, and refiltration if necessary.

AMINES

A variety of amides and ureas are available.

Procedures

ACETAMIDES.   Acetamides of aromatic amines are prepared very readily by dissolving about 0.5 g of the amine in 1 to 2 ml of acetic anhydride and heating for a few minutes. Then add a few ml of water and warm the mixture until the excess acetic anhydride is destroyed (liquid disappears). If the amide does not crystallize directly, it can be extracted with ether and recovered as previously described.

BENZAMIDES.   Mix the amine (1 g) with 10 to 15 ml of 10 per cent sodium hydroxide solution and add 2 ml of benzoyl chloride. Shake or stir the mixture for 10 minutes and isolate the amide by collecting the solid and washing with water. In some cases it may be desirable to add a solvent such as methylene chloride, and recover the amide from solution.

SUBSTITUTED UREAS AND THIOUREAS.   The reaction of an amine with an isocyanate or isothiocyanate leads to the corresponding urea; the precautions about water mentioned under alcohols must be kept in mind with isocyanates. Isothiocyanates are much less reactive; reaction with amines occurs on warming, but hydrolysis with water is negligible and an excess of the reagent can be eliminated by recrystallization.

$$ArN{=}C{=}O + RNH_2 \rightarrow ArNHCONHR$$

$$ArN{=}C{=}S + RNH_2 \rightarrow ArNHCSNHR$$

UREAS.   To 1 g of the amine add 0.5 ml or less of phenyl or α-naphthyl isocyanate; a few ml of methylene chloride or benzene can be used as a diluent. Excess amine should be removed by washing the solution with dilute acid before isolating the urea.

AMIDES AND NITRILES

Alkaline hydrolysis of these compounds leads to the acid salt and ammonia or the free amine.

$$RCONHR + H_2O \xrightarrow{NaOH} RCO_2Na + RNH_2$$

$$RC{\equiv}N + H_2O \xrightarrow{NaOH} RCO_2Na + NH_3$$

Amides or esters of carbamic acids give amines, alcohols, and CO_2:

$$CH_3NHCONH_2 + H_2O \xrightarrow{NaOH} CH_3NH_2 + NH_3 + CO_3^=$$

$$C_6H_5NHCO_2C_2H_5 + H_2O \xrightarrow{NaOH} C_6H_5NH_2 + C_2H_5OH + CO_3^=$$

With high-melting amides or nitriles of aromatic acids, the rate of hydrolysis and the solubility are low, and alcoholic alkali or hydrolysis in acid is recommended. Most nitrogen-containing compounds are quite soluble in 40 to 60 per cent sulfuric acid and fairly high temperatures can be obtained by refluxing such solutions.

$$ArCONHR' + H_2O \xrightarrow{H_2SO_4} ArCO_2H + R'NH_3^+HSO_4^-$$

An additional possibility for the characterization of aromatic nitriles is conversion to the amide. This reaction, which is strictly speaking not a hydrolysis but a hydration, is carried out by treatment of the nitrile with alkaline hydrogen peroxide; hydroperoxide ion (OOH^-) is the specific reagent involved:

$$RC{\equiv}N + 2H_2O_2 \xrightarrow{OH^-} RCONH_2 + H_2O + O_2$$

Procedures

ALKALINE HYDROLYSIS. Add 1 ml of a liquid or 1 g of a solid amide or nitrile to 30 ml of 10% NaOH or KOH solution and reflux the solution for 30 minutes. Test for the presence of a volatile amine by holding a piece of moist indicator paper at the top of the condenser. Cool the solution to room temperature. If an insoluble amine is present, or if the presence of an aliphatic amine of intermediate solubility is suspected from the odor, extract the solution with several 10- to 15-ml portions of ether, wash the ether solution, dry with $MgSO_4$ or K_2CO_3 and evaporate to obtain the amine. A low molecular weight aliphatic amine will probably be lost in this procedure due to solubility in water and volatility during removal of ether.

After removing the amine, if any, acidify the aqueous solution and collect the acid if it is a solid. A low-molecular-weight aliphatic acid can be isolated by thorough extraction, but will usually be of little value as a derivative.

ACID HYDROLYSIS.   Slowly add 5 ml of concentrated sulfuric acid to 10 ml of water. To this solution add 1 g of the amide or nitrile (this procedure will generally be used with high-melting solids). Reflux the solution for 30 to 60 minutes, then cool in ice and dilute with an equal volume of water. If a solid acid separates, this is collected, washed with water, and characterized.

The amine is recovered by making the aqueous solution basic by addition of 10% NaOH. A heavy precipitate of inorganic salts usually separates, and more water must be added. The amine is then recovered as previously described.

CONVERSION OF NITRILE TO AMIDE.   In a 100 ml, round-bottom flask place 1 ml or 1 g of the nitrile, 5 ml of 20% hydrogen peroxide and 1 ml of 6N (20%) NaOH solution. If the nitrile is insoluble, add 5 to 10 ml of ethanol. The reaction should occur exothermically and cooling may be necessary at first. Keep the temperature at 40 to 50° by cooling or gentle warming for 2 hours. Neutralize the solution and concentrate by distillation to remove most of the alcohol. The amide will usually separate from the aqueous solution on chilling; if it does not, extract with methylene chloride.

CARBONYL COMPOUNDS

Semicarbazones and 2,4-dinitrophenylhydrazines are the most generally satisfactory derivatives of simple aldehydes and ketones:

$$RCOR + NH_2NHCONH_2 \rightarrow \begin{matrix} R \\ \diagdown \\ \diagup \\ R \end{matrix} C{=}NNHCONH_2$$

$$+ NH_2NHC_6H_3(NO_2)_2 \rightarrow \begin{matrix} R \\ \diagdown \\ \diagup \\ R \end{matrix} C{=}NNHC_6H_3(NO_2)_2$$

Procedures

SEMICARBAZONES.   Prepare a solution of 0.5 g of semicarbazide hydrochloride and 1 g of sodium acetate in 2 ml of methanol; the large crystals are ground together with a glass rod and finely divided sodium chloride separates. Add about 0.5 ml of the carbonyl compound and allow the solution to stand for 15 to 30 minutes. The derivative usually crystallizes on evaporation of the methanol or addition of a small amount of water.

2,4-DINITROPHENYLHYDRAZONES. Dissolve about 0.2 g of the solid or liquid ketone or aldehyde in 1 ml of ethanol and add dropwise 3 ml of the 2,4-dinitrophenylhydrazine solution prepared as described on page 174.

ESTERS

Saponification of relatively low-molecular-weight esters can be carried out by refluxing with aqueous alkali. A solid acid can readily be isolated simply by precipitation with acid, but isolation of the anhydrous alcohol for further derivatization is difficult on a small scale, and for this purpose, saponification in diethylene glycol is preferable.

A derivative of the acid portion can usually be obtained directly. Some esters can be converted to amides by heating with ammonia or amines. A more general method involves conversion of an aromatic amine to the more reactive anion by treatment with ethylmagnesium bromide, followed by reaction with the ester.

$$ArNH_2 + C_2H_5MgBr \rightarrow ArNH^-Mg^+Br + C_2H_6$$

$$ArNH^-Mg^+Br + RCO_2R' \rightarrow RCONHAr + R'OMgBr$$

Procedures

SAPONIFICATION. In a 10 ml distilling flask place 3 ml of diethylene glycol (bp 244°), 0.5 g of KOH pellets, and 0.5 ml of water. Heat the mixture until the alkali has dissolved and then cool and add 1 to 2 ml of the ester. Heat again until the ester dissolves and then more strongly, distilling the alcohol into a cooled test tube. The potassium salt of the acid may separate as a solid during the reaction.

CONVERSION OF ESTER TO SUBSTITUTED AMIDE. In a 50 ml round-bottom flask, prepare a solution of ethylmagnesium bromide from 10 ml of anhydrous ether, 1 ml of ethyl bromide, and 0.2 g of magnesium (see Chapter 18). After formation of the Grignard compound and brief refluxing to ensure completion, add a solution of 1.5 ml of aniline or 1.5 g of p-toluidine in a little dry ether. After reaction of the Grignard reagent is complete, add 1 ml of the ester and reflux the mixture for 10 minutes. Cool, add 10 ml of 2N HCl and shake to extract unreacted amine, adding more ether if needed. Isolate the derivative as usual from the ether solution.

HALIDES

Alkyl halides that contain no reactive functional groups can be derivatized by conversion to the Grignard compound and treatment of the latter with an aryl isocyanate to give a substituted amide of the homologous acid; this method is quite general for halides that can form Grignard compounds.

$$R—X \xrightarrow{Mg} RMgX \xrightarrow{ArNCO} R—\overset{\overset{\displaystyle O}{\|}}{C}—NHAr$$

Procedure

Convert 1 to 2 ml of the halide (sample must be anhydrous) to the Grignard compound in the usual way (see Chapter 18) with about 0.2 g of magnesium and 10 ml of anhydrous ether; a crystal of iodine can be added if needed. To this solution add, in small portions, a solution of 0.3 to 0.5 ml of the isocyanate (an excess must be avoided) in 5 ml of ether or methylene chloride. After hydrolysis with $2N$ HCl, isolate the amide from the ether solution in the usual way.

NITRO COMPOUNDS

For aromatic nitro compounds oxidation of side chains or further nitration often provides satisfactory derivatives. Another general approach is reduction to the amine, which is usually then converted to an amide.

Procedure

In a 100 ml round-bottom flask with reflux condenser place 2 g of tin (granulated or mossy) and 1 g of the nitro compound. Twenty ml of $2N$ HCl is then added, through the condenser, in small portions with constant shaking. After warming on the steam bath for 10 minutes, cool the solution and slowly add 40% NaOH solution to liberate the amine, which can then be steam distilled or extracted with ether and further characterized. Alternatively the procedure in Chapter 31 can be used.

References

Z. Rapoport, *Handbook of Tables for Organic Compound Identification.* Chemical Rubber Co., Cleveland, 1967.

D. J. Pasto, and C. R. Johnson, *Organic Structure Determination.* Prentice-Hall, Englewood Cliffs, N.J., 1969.

R. L. Shriner, R. C. Fuson, and D. Y. Curtin, *The Systematic Identification of Organic Compounds,* 5th Ed. John Wiley and Sons, New York, 1964.

A. I. Vogel, *A Textbook of Practical Organic Chemistry,* 3rd Ed. Longmans, Green and Co., New York, 1957.

L. F. Fieser, and M. Fieser, *Reagents for Organic Synthesis.* John Wiley and Sons, New York, 1967.

INDEX

229

SELECTED ATOMIC WEIGHTS

Aluminum	26.98	Magnesium	24.31
Barium	137.34	Manganese	54.94
Boron	10.81	Mercury	200.59
Bromine	79.91	Nitrogen	14.01
Calcium	40.08	Oxygen	16.00
Carbon	12.01	Phosphorus	30.97
Chlorine	35.45	Potassium	39.10
Chromium	52.00	Selenium	78.96
Copper	63.54	Silica	28.09
Fluorine	19.00	Silver	107.87
Hydrogen	1.008	Sodium	22.99
Iodine	126.90	Sulfur	32.06
Iron	55.85	Tin	118.69
Lithium	6.94	Zinc	65.37

COMMON ACIDS AND BASES

	moles/liter	sp.gr.	g/100 ml
Hydrochloric acid, concentrated (37%)	12.0	1.19	44.0
, 10%	2.9	1.05	10.5
, 1N	1.0	1.02	3.6
Ammonium hydroxide, concentrated	15	0.90	25.6 (NH_3)
Sodium hydroxide, 10%	2.8	1.11	11.1
Sodium bicarbonate, saturated, 20°	1.1	1.06	9.5
Sodium carbonate, saturated, 20°	2.1	1.19	21
Sulfuric acid, concentrated (96%)	18	1.84	177
Nitric acid, concentrated (71%)	16	1.42	101
Hydrobromic acid, concentrated (48%)	8.8	1.49	71
Hydriodic acid, concentrated (57%)	7.6	1.7	97
Acetic acid, glacial	15.9	1.05	105